The Charm Of Creation

A Devotional Study

A ROMAN CONSUL

The Charm Of Creation

A Devotional Study

ForeWord by Lowell G. Dooley,

Joseph Kennedy

Thee Alabaster Box
Southside, Alabama

We want to hear from you. Please send your comments about this book to us at the address below. Thanks you.

Thee Alabaster Box
2106 Keenum Drive
Southside, Alabama 35907
phone 256 442 1466
email jtk6@bellsouth.net

Books by Joseph Kennedy

Onesimus: Flight to Paradise
Lois: the Beauty of Holiness
The Invalid Warrior: Life of Paul
Chained to His Chariot: the Faithful Warrior
The Latter Years: the Ageless Warrior
From Eden to Canaan I: How the World Came to be the Way it is
From Eden to Canaan II: How the World Came to be the way it is
Epistle to the Hebrews
The Charm of Creation
Knight of the Hour Glass
Reverie: Poems to Ponder

Foreword

Being raised in a Christian home, I do not remember a time that I doubted that "In the beginning, God created..." I just accepted the fact that God is who He says he is and God did what He said He did. I did not question His sovereignty and power.

I spent my formative educational years during the 1970's-1980's. not only did this includee primary and secondary education, but also three earned degrees in healthcare along with a specialization certificate. I was exposed to both creationism, and evolutionism.

During my professional educational process, I began to believe the evolutionary teachings. I did not eliminate the creative aspect of God. I did not doubt God. I just tried to marry the two together. It made perfect sense to me that God created the universe and then allowed evolutionary processes to take over – more of a deistic approach to creation. God spoke the universe into existence and then allowed it to develop on its own. This caused quite a spiritual struggle within me – often going unrecognized.

Ten years ago, God allowed me to attend a service where Dr. Joseph (Dr. Joe) Kennedy was preaching. His topic was creationism. During this service, the scriptures came alive. My creationism "light bulb" clicked on. I began to understand that God describes exactly His creative work, and that He was intimately involved in the entire process – 6 days/ 24 hours.

Now I understand the importance of taking God's Word literally. He created the entire universe "ex nihilo" (from nothing) in 6 twenty-four periods. He spoke everything into existence except man – He personally formed man from clay. What a beautiful picture of a loving God and His creation.

I also understand that the Accuser is constantly trying to cast doubt over this creative process. He knows that if man would doubt the accuracy of Genesis 1-3, then he would more easily doubt his need for a Redeemer. If he would doubt his need for a Redeemer, then fellowshp between the Creator and the pinnacle of His creation, man, is broken.

I am thankful that Gode placed Dr. Joe in my path during my struggles with creationism. He is a dear friend, and my mentor. Through Dr. Joe, God has ignited a passion for creationism and calmed my restless soul. With great honor and joy, I recommend his writings. God has used Dr. Joe to reveal His spiritual truths.

Lowell G. Dooley, DRNA, BSN, MS
October 14, 2009

Contents

Book I
Created Conscience

Introduction to Book I   **7**

1. Two Natures	Romans 7:15-25	17
2. David's Conscience	I Samuel 24	21
3. A Good Conscience	Acts 24:16	25
4. The Pricked Conscience	Acts 2:37	29
5. A Guilty Conscience	Psalm 32	33
6. The Conscience	Romans 2:14-15	37
7. Joseph's Brothers	Genesis 50:17	41
8. Work of the Conscience	Romans 2:14	45
9. Evolutionist's Conscience	I Timothy 4:2	49
10. The National Conscience	Psalm 9:17	53
11. Word Study	John 8:9	57
12. Northern Kentucky	Titus 1:15	61
13. The Testimony of Conscience	I Corinthians 1:12	65
14. My Brother's Conscience	I Corinthians 8	69

Book II
Creative Miracles

Introduction to Book II   **73**

15. Creative Miracles	John 2:7-11	75
16. Created Manna	Exodus 16:4	79
17. Created Bread	John 6:1-14	83
18. Created Human Tissue	John 11:1-44	87
19. A Created Coin	Matthew 17:34-27	91
20. Created Signs	John 20:30:31	95
21. Healing a Mother in Law	Maark 1:29-31	99
22. Created Nerves	Mark 5:24	103
23. Made Whole	Matthew 15:28	107
24. Signs	Matthew 12:39	111
25. Created Anti-gravity	John 6:16-21	115
26. Created Oil	I Kings 17:8-16	119
27. Created Brimstone	Genesis 19	123
28. Made Flesh	John 1:14	127
29. The Holy Saviour	Luke 1:36	131

Book III
Creative Psalms

Introduction to Book III   135
Brimstone   136
30. The Ungodly   Ephesians 3:9   137
31. Heavenly Glory   Psalm 19   141
32. The Voice of GOD   Psalm 29   145
33. Absolute Justice   Psalm 9   149
34. GOD's Possessions   Psalm 2:1   153
35. The Breath of GOD   Psalm 33:6-9   157
36. A Clean Heart   Psalm 51:10   161
37. Wonderful Works   Psalm 40:4   165
38. The Everlasting GOD   Psalm 90:1,2,4   169
39. Stormy Seas   Psalm 107:23-28   173
40. Man's Earth   Psalm 115:16   177
41. A Sure Foundation   Psalm139   181

Book IV
Creation in the New Testament

Introduction to Book IV   185
Psalms   186
42. Matthew and Genesis   Matthew 19:5   187
43. Male and Female   Mark 10:6   191
44. Treading on Serpents   Luke 10;19   195
45. The Father of Lies   John 8:44   199
46. The Maker of all Things   Acts 14:15; 17:24   203
47. Invisible Things   Romans 1:20   207
48. Man-made Fossils   I Corinthians 15:2   211
49. The Woman's Seed   Galatians 4:4   215
50. The Fellowship of Mystery Ephesians 3:9   219
51. Knowledge   Colossians 3:10   223
52. The Beauty of Work   I Thissalonians 3:2   227
53. Good Creatures   I Tinothy 4:4   231
54. Grace in Eternity   II Timothy 1:9   235
55. God's Work   II Timothy 3:16   239
56. Hope of Eternal Life   Titus 1:2   243
To Darwin   248
57. God's Rest   Hebrews 4:10   249
58. Downward Change   II Peter 3:4   253
59. Explaining the Unexplainable I John 3:8   257
60. No More Time   Revelation 10:6   261
Index   267

Introduction

Science and the technology it makes possible have almost destroyed the earth. Science and technology cannot be blamed, however, for it is mankind's use of them that has damaged the earth. Mankind has used science and technology wrong because man himself is wrong. He is wrong in ignoring the will of God. It seems that in six thousand years humanity would have learned that God's ways are best. Men who have power over the earth seem to be the most rebellious of men. These people make animals more important that the souls and bodies of men.

Man has used his divinely provided brains to rebel again God, and to reverse every one of God's orders. God said that men and women were to reproduce upon the earth. Man set out to invent chemicals that prevent women from conceiving. When they do conceive, many use modern technology to destroy the baby. God said man would earn his bread by the sweat of his brow, and man invented tractor cabs that have all the amenities that make labor easy and pleasant. God said woman's desire would be to her husband, and women declared themselves free. God commanded man to care for the earth and to dress it, and man makes the earth as ugly and unproductive as possible.

A Law of science never changes. God created these Laws when He created matter. No discovery man can make, and no idea that he formulates can be correct if it cannot pass the test of the Laws of physics. Man is very limited in his discoveries. Man can move out into space a short distance, but the immensity of the universe is unmanageable for man, because man must breathe, and he can only live seventy or eighty years. All that God wants man to know about the universe, He has written in His book.

Man can know more about the origin of the earth by faith, than he can by digging in the earth. There is more information to be had from the pages of the Bible that from other sources. A man on TV turned over a stone on an Australian desert. He said, There, the sunlight struck that place under that stone for the first time in millions of years. This statement was made by a scientist, and shown on a serious TV channel. The statement was silly, and should have been taken as truth by nobody. To be a successful paleontologist, one has to be a good yarn-spinner. Myth making is the stuff of much scientific study.

God alone is able to work a miracle. I believe a miracle is the reversal or suspension of the Laws of nature. Nearly all miracles required the power of creation. The LORD has reserved to Himself the power of creation. When we say a man creates something, all he has actually done is put together certain elements that are easily available. A painter does not create a painting – he only smears matter on matter that looks good. I heard a man addressing a crowd in a church in Minsk, Belarus one night. He boasted at length about the healing he had done in other places. He had caused the pupils of a child's eyes to appear where there had been none. He had done wonderful deeds. He asked people to come forward who needed to be healed, and he put on a show of healing. I believed none of it. What a waste of time, I thought, when Jesus Christ could have been lifted up for needy people. Religious charlatans abound today, even in our so-called scientific world.

What can a man write today? Mark Twain and Will Rogers have already made all the good cracks about politicians. Winston Churchhill and Ronald Reagan have made all the witty remarks about political matters. Books abound about every subject in the Bible. If Dr. M. R. DeHann or Dr. William Henry Green or Dr. Henry M. Morris were alive, they would have no problem writing a good book. Can there be too many books written about creation? Evidently, publishers think so. John said that if all the works of Jesus were written in books, the world could not contain them. If all the books that could be written about creation were written, they would likely fill up the space between here and the moon. So I am going to add my two cents' worth, and enjoy the labour. I leave them to the world with my love. May God be praised in every Word. The world will be without excuse for not believing.

You may be interested in this bit of information about the writers' of the psalms:

Introduction to Book I

The following are all the scripture references to the Word "**conscience**" in the Bible. Notice that they are all found in the New Testament.

John 8:9: " *And they which heard it, being convicted by their own conscience, went out one by one, beginning at the eldest, even unto the last: and Jesus was left alone, and the woman standing in the midst."*

Acts 23:1: *"And Paul, earnestly beholding the council, said, Men and brethren, I have lived in all good conscience before God until this day."*

Acts 24:16: *"And herein do I exercise myself, to have always a conscience void of offence toward God, and toward men."*

Romans 2:15: *"Which shew the work of the Law written in their hearts, their conscience also bearing witness, and their thoughts the mean while accusing or else excusing one another;)"*

Romans 9:1: " *I say the truth in Christ, I lie not, my conscience also bearing me witness in the Holy Ghost,"*

Romans 13:5: *"Wherefore ye must needs be subject, not only for wrath, but also for conscience sake."*

I Corinthians 8:7: " *Howbeit there is not in every man that knowledge: for some with conscience of the idol unto this hour eat it as a thing offered unto an idol; and their conscience being weak is defiled."*

I Corinthians 8:10: *"For if any man see thee which hast knowledge sit at meat in the idol's temple, shall not the conscience of him which is weak be emboldened to eat those things which are offered to idols; offered to idols;"*

I Corinthians 8:12: *"But when ye sin so against the brethren, and wound their weak conscience, ye sin against Christ"*.

I Corinthians 10:25 *"Whatsoever is sold in the shambles, that eat, asking no question for conscience sake:"*

I Corinthians 10:27: *"If any of them that believe not bid you to a feast, and ye be disposed to go whatsoever is set before you, eat, asking no question for conscience sake."*

I Corinthians 10:28: *"But if any man say unto you, This is offered in sacrifice unto idols, eat not for his sake that showed it, and for conscience sake: for the earth is the LORD's, and the fulness thereof."*

I Corinthians 10:29: *"Conscience, I say, not thine own, but of the other for why is my liberty judged of another man's conscience?"*

II Corinthians 1:12: *"For our rejoicing is this, the testimony of our conscience, that in simplicity and godly sincerity, not with fleshly wisdom, but by the grace of God, we have had our conversation in the world, and more abundantly to you-ward"*

II Corinthians 4:2: *"But have renounced the hidden things of dishonesty, not walking in craftiness, nor handling the Word of God deceitfully; but by manifestation of the truth commending ourselves to every man's conscience in the sight of God."*

I Timothy 1:5: *"Now the end of the commandment is charity out of a pure heart, and of a good conscience, and of faith unfeigned:"*

I Timothy 1:19: *"Holding faith, and a good conscience; which some having put away concerning faith have made shipwreck:"*

I Timothy 3:9: *"Holding the mystery of the faith in a pure conscience."*

I Timothy 4:2: *"Speaking lies in hypocrisy; having their conscience seared with a hot iron;"*

II Timothy 1:3: *"I thank God, whom I serve from my forefathers with pure conscience, that without ceasing"*
have remembrance of thee in my prayers night and day;

Titus 1:15: *Unto the pure all things are pure: but unto them that are defiled and unbelieving is nothing pure; but even their mind and conscience is defiled*

Hebrews 9:9: *Which was a figure for the time then present, in which were offered both gifts and sacrifices, that could not make him that did the service perfect, as pertaining to the conscience;*

Hebrews 9:14: *How much more shall the blood of Christ, who through the eternal Spirit offered himself without spot to God, purge your*
conscience from dead works to serve the living God? **Hebrews 10:2**: *For then would they not have ceased to be offered? because that the worshippers once purged should have had no more conscience of sins.*

Hebrews 10:22: *Let us draw near with a true heart in full assurance of faith, having our hearts sprinkled from an evil conscience, and our bodies washed with pure water.*
conscience from dead works to serve the living God? **Hebrews 10:2**: *For then would they not have ceased to be offered? because that the worshippers once purged should have had no more conscience of sins.*

Hebrews 10:22: *Let us draw near with a true heart in full assurance of faith, having our hearts sprinkled from an evil conscience, and our bodies washed with pure water.*
conscience from dead works to serve the living God? **Hebrews 10:2**: *For then would they not have ceased to be offered? because*

that the worshippers once purged should have had no more conscience of sins.

Hebrews 10:22: *Let us draw near with a true heart in full assurance of faith, having our hearts sprinkled from an evil conscience, and our bodies washed with pure water.*

"**CONSCIENCE - a person's inner awareness of conforming to the will of God or departing from it, resulting in either a sense of approval or** condemnation.

"An inner moral sense within man which condemns or approves his conduct. The Word is a Greek Word and the concept does not appear in the OT (though there is a sense of guilt expressed, Gen. 3:8, 42:21). Paul is the principal NT user (elsewhere only in Paul's speeches in Acts, 1 Peter, and Hebrews)."

The New Smith's Bible Dictionary, William Smith, pg, 71.

Blessed is a tender conscience. Few things are more important to humans. A tender conscience will bring a child to Christ. That is one reason it is so important for a child to submit to Jesus Christ. As people grow older, the conscience tends to fade. I remember when my own throat would seem to close up when I was smitten by my conscience. That is one of the factors involved in my young conversion. Even those who don't seem to have any feelings, have a remnant of conscience. May God be praised for conscience.

Book I

ROMANS 7:15 – 25

TWO NATURES

The dictionary says the conscience is "1. The faculty of recognizing the distinction between right and wrong in regard to one's own conduct. 2. Conformity to one's own sense of right conduct."

"And they which heard it, being convicted by their own conscience, went out one by one, beginning at the eldest, even unto the last: and Jesus was left alone, and the woman standing in the midst." (John 8:9)

According to the news I heard this morning, Life magazine will publish some pictures in the November issue that were made by a well- known photographer. I commend this man for his skill as a cameraman, and his knowledge of the available technology in this field. No doubt he is an expert. The pictures are of fetuses, which I understand, - 17 -chickens and humans, and other creatures. They did look similar - sort of like an apple and a pear are similar. Some of the pictures were shown on tv, and you could see the similarities.

The photographer told about drawings of fetuses that were made many years ago that showed similarities between the fetuses of different animals and humans. He said he wanted to show by photography how the fetuses of different animals looked similar just as the scientist had showed them in drawings years ago. It has always interested me, in an aggravating sort of way, that some people take more pride in thinking of themselves as animals rather than creatures made in the likeness of the Creator.

The photographer did not give any more details, but I know what he was referring to. I have discussed Ernst Haekel (Hay' kel) more than once in my books. He was the man who produced the drawings the photographer referred to. Haekle was a rabid evolutionist who falsified a great deal of things in his efforts to prove evolution. His contemporaries called him the mad man of Jeno University. The drawings referred to on the tvthis morning were Haekel's drawings.

The purpose of all this is not to show that all creatures are similar because they were designed and created by a Supreme Creator, but to try to show that they all had a common ancestor, which was a single cell in some primeval swamp. Haekel's notion, called the recapitulation theory, has long ago been shown to have nothing to do with evolution, and neither do these photographs in *Life* magazine have anything to do with evolution.

Science has advanced since Ernst Haekel so that we can now know more about what is in the cell that Haekle, and Charles Darwin, and others in the 19th century knew only as mysterious "black boxes." No longer must scientists simply look at the outward appearance of an organism and make judgments about it. We understand more about genetics and the complicated DNA molecule, and other factors controlling life. It is in these entities that scientists discover that evolution could not have occurred. Genes control and limit the ability of an organism to differ from one individual to another, and the gene makes the offspring the same kind as its parent.

I have three parents. My earthly mother and daddy brought me into the world in 1927. My body was made what it is by the union of the genes of my parents. I was 100% human. There was not one cell in my body that was anything but human. There was not one cell in my body that could have been some higher form of life than my parents. I was exactly what they were - a human being. My parents also gave me a sin nature. They were born sinners just as their parents were, and so on back to Adam. In fact, I was in Adam when God created him, just as my parents were, and so I sinned in Adam.

I am a human sinner. I was born a sinner, and then by the time I was six years old, I had chosen to be a sinner. I was taught by faithful Sunday school teachers and my pastor, that Jesus, God's only begotten Son, died on the cross of Calvary for me to save me from sin. That was all I knew. I hated the bad things I was doing, and so I trusted Jesus to save me, and He did. That was all I knew at that time. At that moment I was born again. I now had the third parent, God; the Father was now my Father. From Him I received a new nature. This nature hated sin, and did not sin. My new nature has never sinned. It never will. My old nature did not die- the Adamic nature I received from my parents - did not die or go away. It has tormented me all my life. It will torment me until I die. That old nature has been the target of Satan's darts for 62 years. It is now old and weak, but it still responds when the devil prods it. My old nature would keep me from praying. My old nature would keep me from singing hymns. My old nature would go fishing on Sunday morning. My old nature would keep all my money and never put a dime in church. My old nature is almost as bad as Satan himself. I hate it with a passion. I would kill it and throw it into a ditch if I could. I crucify it every day, but it won't stay dead.

My new nature loves to go to church. My new nature I received from my heavenly Father loves to sing hymns. My new nature loves the Word of God. My new nature loves the souls of men, and weeps for them. My new nature forgives people when they tramp on my feelings or ignore me. My new nature would give every penny I can rake and scrape to the LORD. My new nature will never die. When this robe of flesh gasps its last breath, my new nature will be free at last.

There is a warfare between my old nature and my new nature that rages like a tempest within me. Sometimes they pull me apart. **Paul** said it better than I could say it. Listen:

"For that which I do I allow not: for what I would, that do I not; but what I hate, that do I. If then I do that which I would not, I consent unto the Law that it is good. Now then it is no more I that do it, but sin that dwelleth in me. For I know that in me (that is, in my flesh,) dwelleth no good thing: for to will is present with me: but how to perform that which is good I find not. For the good that I would I do not: but the evil which I would not, that I do. Now if I do that I would not, it is no more I that do it, but sin that dwelleth in me. I find then a Law, that, when I would do good, evil is present with me. For I delight in the Law of God after the inward man: But I see another Law in my members, warring against the Law of my mind, and bringing me into captivity to the Law of sin which is in my members. O wretched man that I am! who shall deliver me from the body of this death? I thank God through Jesus Christ our LORD. So then with the

mind I myself serve the Law of God; but with the flesh the Law of sin."

Paul sounded like a crazy, mixed up kid, but he was as sound as a dollar because he understood what was going on within him. Psychologists steer people away from the Gospel, the Bible, and the church, telling them to do their own thing. But sinners have a conscience, and until that conscience is defiled it will cause misery and unhappiness if the person engages in sin. Conscience will be happy if the person goes to church once in a while, like on Christmas or Easter, or to a funeral. Conscience is easy to put to bed. A few dollars to the United Way will quiet the conscience. The conscience is satisfied with an occasional glance at the Word of God. Paul is not talking about his conscience.

The conscience can make a person a better citizen - a better person - but it will not take him to Heaven. The new man is not our conscience. The new man loves to go to church. The new man loves the things of God and His Word. A saved person will go to church because He loves it, and because he knows it pleases God. A great number of sinners think they are saved because they do enough righteousness to keep their conscience quiet.

Many sinners put Saints to shame by the right conduct of their daily life. Many sinners are honest and kind. Their behavior is exemplary. Many sinners are humanitarians, generous in their gifts to the needy. The most gracious sinner is guilty of the sin that condemns a human to everlasting fire. Only one sin condemns, remember that. Refusing to give Jesus Christ His rightful place. Refusing to acknowledge that Jesus Christ came in the flesh. A Saint is similar to a sinner in many ways. A Saint goes to restaurants to eat sometimes. Saints drive cars. Saints write checks. Saints watch tv sometimes. Saints do a lot of things that sinners do. Saints even eat. The Saint has heavenly genes.

DAVID'S CONSCIENCE

In the last essay, what came to me so stunningly was the truth that conscience can be a major factor in causing people to think they are saved, when in fact, they are not. We must be very careful that we do not think that just because we feel righteous that we are saved. When Isaiah saw the LORD in that great vision in chapter 6, his response was *"Woe is me! for I am undone; because I am a man of unclean lips, and I dwell in the midst of a people of unclean lips; . . . ."*.

A close walk with the LORD will reveal to us our own sinful nature. We become aware not only that we are sinners, but at the same time, have the assurance that the blood of Jesus Christ has cleansed us from all our sins.

Let's go to the dictionary to learn what the Word "conscience" means. My dictionary says, "n. 1. the sense of right and wrong; ideas and feelings within a person that tell him when he is doing right and warn him of what is wrong."[1]

The conscience is formulated by the experiences of the individual, and so a conscience in Alaska can be different from a conscience in the southern U.S.A. A headhunter in New Guinea would have a different conscience from a lobster fisherman in New England because they grew up in different cultures.

Paul sounded like a crazy, mixed up kid in Romans 7:15-25, but he was as sound as a dollar because he understood what was going on within him. But sinners have a conscience (Romans 2:14-16), and until that conscience is defiled (Titus 1:15) or seared (I Timothy 4:2) it will convict the sinner of violating it.

Man has a sense of right and wrong because he is a created being, made in the likeness and image of God, and therefore has a consciousness of the existence of God. However let it understood that the conscience can culturally formed, and therefore cannot be depended upon fully as a guide to right and wrong. A headhunter has no feeling of wrong when he is eating the flesh of another human being. A very old woman was unable to eat, and she made the remark to the missionary doctor that she believed she could take a little food if she had the hand of a small boy cooked well done. She had no conscience about eating human flesh.

Cultures that practiced human sacrifice had no feeling of wrong when they captured an enemy and brought him to the priest to be offered to their god as a sacrifice. People raised in a home where God's name is taken in vain as a regular part of the conversation have no conscience about taking God's name in vain. Sinners do have a conscience, but it is not to be trusted. The old saying, "Let your conscience by your guide" was made up by somebody who did not understand the conscience very well. God gave man a conscience to aid him in making day to day decisions, but after man became a sinner, the conscience could no longer be trusted

In I Samuel 24 we find King Saul pursuing David with the intention of killing him. Saul went into a cave with his small army to spend the night. David and his band of men had already taken refuge in this cave, but in the **darkness**, Saul and his men did not know David and his men were in there hiding in crevices of the cave walls. When Saul lay down to sleep, David's men encouraged him to go to Saul, and kill him. David quietly crept to Saul's side, and cut off the skirt of his robe. He could not bring himself to kill a man who was God's anointed (see 25:6). The Bible says in 4 and 5,

". . .*Then David arose, and cut off the skirt of Saul's robe privily. And it came to pass afterward, that David's heart smote him, because he had cut off Saul's skirt.*"

David had a conscience attack. It did not matter that he had spared the life of his enemy. A little bit of good mixed with wrong will not satisfy a good conscience. His conscience smote him. **David**'s conscience smote him because he had a heart because he was a man who knew the LORD God. The conscience can only attack a tender heart. David was a warrior who was a fearless leader of some of the greatest warriors who ever lived. David struck down Goliath, and cut off his head without a twinge of conscience.

In the gray light of morning, Saul arose to continue his pursuit of David, but David called out to him, and told him that he had spared his life during the night. Saul recognized that David was more righteous than he, but we are not told that Saul's heart smote Yes, he made a fine speech about how David had showed him mercy, but there is no indication that he had a conscience. David took an oath that he would spare Saul's family, an oath which he kept when he at last became king, but there was no promise from Saul that he would not continue his efforts to kill David. Saul tried to kill David until the day he died.

We can never say anything absolutely about what goes on inside an animal's head or heart, because we cannot know what is there, but as far as we know, an animal does not have a conscience. It has never been scientifically demonstrated that an animal has a conscience. An animal has no moral sense of what is right and wrong. A cat will play with a mouse until the mouse dies of exhaustion or injuries, but the cat will never feel sorry for the mouse. A dog will chase an animal and kill it just for sport, and never feel sorry for it. Most human hunters do have some feeling for the animals they kill as indicated by their effort to kill the animal as painlessly as possible.

How did it come to pass that humans have a conscience, and animals do not? Could the conscience evolve? How did matter become sensitive to wrong to such an extent that man's conscience will hit him like hitting him with a hammer. I have felt conscience pains so strongly that it felt like a knife through my throat. You may have had the same experience.

Our conscience, though, is not a completely reliable guide to righteousness. We cannot depend upon our conscience because over the years we make it do what we want it to do. The conscience of Americans has radically changed in the last few years. The more strongly people declare right to be wrong, and wrong to be right, the less the conscience can be trusted. David was a man with a tender heart, loving the LORD his God with all his mind, body soul, and spirit. Saul was a man who had turned away from God, and had come to worship his crown more than the One Who gave him the crown.

To **David's** conscience, killing the LORD's anointed was such a wrong that he was willing to risk his life to follow his conscience. Saul's conscience was dead, killed by Saul's backslidden condition. **David** was a man of prayer and praise. Saul was a man with a mind deranged by a heart grown cold and indifferent to the presence of his God.

Saul could neglect his responsibilities to God and never feel his heart smite him. Saul could skip prayer time, and never feel his heart ache. Saul could chase an innocent man with the intention of killing him, and never have a heart ache. Saul could turn from God's priests to the devil's witch, and not have a conscience attack.

A saved person cannot neglect the Word of God without his heart smiting him. A saved person cannot refuse to attend church regularly without having a troubled conscience. A saved person cannot live a worldly, sinful life, and not have his conscience tell him he is doing wrong. How about your life, dear reader? Would your pastor recognize you if you attended church Sunday morning? Do you think you could get that teen-aged daughter to go to church with you Sunday morning? Or would she mock you as Lot's sons-in-Law mocked him? No one can be happy with a heart as hard and cold as a stone. That's not normal for a human. A human is born with a conscience, and it is normal to live with a happy conscience. **David** was a happier man than Saul was. Saul was a miserable individual, but **David** was the sweet singer of Israel.

———
1. Clarence L. Barnhart, editor, ("The World Book Encyclopedia Dictionary": Chicago, 1963). p. 423.

A GOOD CONSCIENCE

D r. Luke says, *"And herein do I exercise myself, to have always a conscience void of offense toward God, and toward man."*

"Let your conscience be your guide" is the counsel some people give to others, but is it wise to allow your conscience to be you guide? The Word conscience, according to our dictionary, quoted the other day, is the knowledge of right and wrong. The Word conscience is derived from the Latin Word "conscienta", which is a compound of the prepositions con and "scio", meaning "to know together", or "joint knowledge with others," or "the knowledge we share with another." Conscience is knowledge we share with the LORD. The knowledge of what is right and wrong.

Conscience has the same root Word as "conscious," which means "awareness of." Conscience is an awareness restricted to the moral sphere. It is a moral awareness. The Greek equivalent of the Word is a Word that means "together," "to know." The Word "conscience" is not in the Old Testament, but the idea is well known and expressed by the term "heart." In Job 27:6, Job declares,

"My righteousness I hold fast, and will not let it go: my heart shall not reproach me so long as I live."

Job knew that as long as he behaved righteously that his heart would not reproach him. However, God awakened his conscience to some things he did not realize when He preached to him in the last chapters of **Job,** for **Job** did have a conscience attack, and repented. (Job 42:6)

In Psalm 32:1-5 and Psalm 51:1-9 we read the confessions of a man whose conscience is plaguing him.

The Greeks and Romans personified conscience and depicted it as fiendish female demons called Erinyes and Furies.

In the story of **Joseph** and his brothers, we see the conscience at work in the real world. In chapter 37 of Genesis the story is told of **Joseph**'s ten brothers selling him into slavery, and the tale they told to their father to cover up their crime. There is little indication that these boys had any problem with their conscience at that time. They were glad to be rid of **Joseph**, and though their consciences may have bothered them, nothing of their emotions is revealed at this time. When they visited **Joseph** years later used their consciences to bring them to repentance.

But the years passed, and their aged daddy, **Jacob**, came to the time when he would pull his feet up into his bed, and go to sleep with his fathers. And then fear brought the brothers' conscience back to life, and they were afraid. Genesis 50:15-21 records what happened then. The brothers came to **Joseph** pleading for forgiveness, and as we would expect, **Joseph** was magnanimous with them, and forgave them.

Conscience smites us with a **guilt complex** that can be very harmful if it is not taken care of. When a person has a guilt complex and goes to a psychiatrist or psychologist, the physician will do everything he can to assuage that guilt feeling. He will analyze all the events in the patient's life, and try to find the cause of the guilt. Of course, the patient knows what is wrong, but he is not willing to do what he knows is right to heal his guilt sickness. The psychologist will help him to sear his conscience so that it won't hurt him anymore, and he goes away thinking all is well.

A guilt complex is a very good thing if it is recognized for what it is. When our conscience smites us, then we must put into operation the repentance - confession - forgiveness mechanism. It is very simple and effective. When a wrong is done, there must always be repentance first. That is simply an agreement with God that we have done wrong. Remember, conscience is "a knowledge we share with others," meaning the Father, the Son, and the Holy Spirit, and the Word of God. The knowledge we share is that what we did was wrong. After repentance, there must be confession. Some people would rather die than to confess that they are wrong about something. Many people would rather pay a psychologist thousands of dollars for psychological Band-Aids, than to get the matter right with God and man.

After confession, of course, there must be forgiveness on the part of the person who has been injured by the wrong, providing there was someone injured by the wrong. The wronged person is obligated to forgive, for the Bible commands us to forgive without making any exceptions. We forgive everything that forgiveness is asked for, and as many times as forgiveness is asked for.

After the guilty person makes confession, and asks for forgiveness, his conscience can be at ease, and stop smiting him. It doesn't matter whether the injured person forgives or not, the guilty one has discharged his responsibilities, and can put the matter behind him. He is now back in fellowship with God, and with his conscience.

But what about the injured person? The injured person must forgive, or he takes upon himself the guilt of the situation. If he refuses to forgive as the Bible commands, then he becomes the guilty party, and his conscience should smite him, if he has a conscience. Some people have seared their conscience, and no longer have one.

Everybody at some time or another will have a conscience attack. It is not pleasant. I have lain awake at night staring at the black ceiling because I had committed a crime against God and man. I have had to apologize many times, even to my sons. I have long since come to the place where I can repent and confess without too much pain in my self esteem. All of us sin. All of us offend others. Much of experience sins against us. We should keep our consciences fresh and clear by heeding its warnings, and getting things right again.

Our conscience came from the garden of Eden. Adam and Eve were innocent. They knew not what good and evil were. They had no conscience. They were like little babies who would just as soon run around with no diaper as with one. We can't know how they thought, or what it was like to be so happy and innocent that we did not realize even what good was. Everything for them was perfect, but they did not know it was perfect. They were just perfectly happy.

When they sinned, then they came to know good and evil, for they had eaten of that fruit, and now they must understand evil. Now they would have to contend with a conscience. They had to learn to live with a moral awareness. Man ever since has had to live with a conscience that he will either heed and obey, or he will ignore, and sear with a hot iron. In some nations, the people have been out of touch with God for so long that they have developed consciences that cannot be trusted. Here in America, there are great numbers of people who have no godly moral awareness. In some families, it is the norm to steal, to defraud. This is the way they have been taught to believe.

The first time the Word conscience is used in the New Testament is in John 8:9 where the scribes and Pharisees brought before Christ a woman that was taken in adultery. The story is well known among Christians. This is the event when the Savior stooped down and wrote on the ground. Probably, he wrote some of the Ten Commandments. Then He rose up and said,

". . .He that is without sin among you, Let him first cast a stone at her."

He stooped and wrote again on the ground, and when he rose up the second time, there were no accusers in sight. They had all disappeared.

Verse 9 says,

"And they which heard it, being convicted by their own conscience, went out one by one, beginning at the eldest, even unto the last: and Jesus was left alone, and the woman standing in the midst."

Strive earnestly *"... to have always a good conscience void of offence toward God, and toward men."* is something that all of us should strive for.

THE PRICKED CONSCIENCE

It is interesting, but a little bit distressing, how Christians use Words that are not in the Bible. The Word "conviction" is one of the main Words in our vocabulary. I have used it all my life to describe the emotional state of a person who has had his conscience pricked by the preaching of the Word of God. The Word "conviction" is not in the Bible, and the Word "convicted" is in the Bible one time. I will have to make another change in my terminology in order to say what the Bible says in the way the Bible says it.

Let's see today just how the Bible describes this condition we call conviction. The first instance of people being smitten in their conscience by the preaching of the Word of God is found in Acts 2:37. Peter has preached the great sermon on the day of Pentecost when ". . .*about three thousand souls. . .*" were added to them, and had just spoken the Words recorded in verse 36,

"Therefore let all the house of Israel know assuredly, that God hath made that same Jesus, whom ye have crucified, both LORD and Christ.",.. when those who heard "...*were pricked in their heart ... ."*

This terminology is completely in keeping with what we are told the Bible is and does.

"For the Word of God is quick, and powerful, and sharper than any two-edged sWord, piercing even to the dividing asunder of soul, and spirit, and of the joints and marrow, and is a discerner of the thoughts and intents of the heart."

The Word of God pricks or stabs the conscience, and causes the conscience to attack the sinner so powerfully that he must do something to alleviate the pain. Our conscience hits us repeatedly and without letup for a long time. The Word of God brings a person into conflict with his conscience, and the conflict lasts until the person either makes his conscience good by repentance, confession, and faith in Christ; or he sears it, or defiles it, or weakens it, and makes it of no effect.

The preaching of the Word of God is designed to awaken the conscience, the person's moral awareness. In Acts 23:1 **Paul** said,

"*... **Men and brethren, I have lived in all good conscience before God until this day.**"*

Paul's testimony is that he had enjoyed a good conscience before God all his life. That is the statement that caused the high priest to command him to be struck across the mouth, a violation of the Law of Moses. That interruption may have interfered with **Paul** giving further explanation to a very difficult problem here. He repeats this about his conscience in his speech before Felix in 24:16.

If **Paul** had lived in all good conscience until that day, that would include the days during which he persecuted the church. We examine his testimony further in Romans 7:7-14, and discover that he had been guilty of

"*... **all manner of concupiscence.**"*

which is sin. Sin rouses the conscience which brings the sinning one to repentance, and the conscience is again made good. In verse 11 of Romans 7 **Paul** confesses that sin had deceived him by taking occasion by the commandment.

This illustrates the importance of having a conscience that is programmed by the Gospel of Jesus Christ and the New Testament. Under the Law of Moses, **Paul** thought he was a good man, and he was because he obeyed the letter of the Law, but when he realized that the Law had been fulfilled in Christ, his conscience had to be reprogrammed. No longer was it Lawful for a man to put people to death because they rejected the Law of Moses. **Paul** had indeed had a clear conscience before God, but his conscience was based upon principles that God changed, and therefore his conscience had to change to be good under the new covenant.

Paul's testimony in Philippians 3 reveals that he had reason to be confident in the flesh, for he was blameless as touching the righteousness which is in the Law. However, he continued to trust in his righteousness in the Law after the Law had been disannulled, and so was deceived during that time for no man can be righteous under the Law of works after the resurrection of Jesus Christ.

Paul's conscience was always good before the LORD, but that teaches us that we cannot trust our conscience unless it is geared to the truth. There are vast numbers of people today who have consciences that are good, but that is because they are deceived, and their conscience cannot function as a reliable guide.

A lost person can do enough good things to keep his conscience from smiting him. A lost church member will go to church several times a year, not because he loves to go to church, but because it makes him feel good, and his conscience leaves him alone. People do all sorts of good works in an effort to have a good conscience, and they succeed in deceiving themselves and their conscience.

It requires potent preaching from the Word of God to arouse the conscience of such persons so that their conscience will smite them. Those three thousand people that got saved on the day of Pentecost were all devout Jews with no conscience problem. They were perfectly content with their rituals and ceremonies according to the Law of Moses. Many of them were Jerusalem Jews, living in close proximity to the temple. Their conscience did not trouble them. They thought they were as righteous as Moses himself. But they were deceived, and their conscience was dead.

The Gospel of the death, burial, and resurrection of David's divine Son pricked them in their heart, and their conscience smote them with such pain that they cried out,

"Men and brethren, what shall we do?"

How can we get this sWord out of our heart? The most sinister element of sin is that it conceals the awfulness of sin. The sinner thinks he is a great guy. He does not recognize the evil of rejecting his Creator, and refusing to believe the Gospel. In fact, he may confess all that, yet feel he is so good he was not included in the murder of the Son of God. He does so many good things that his conscience is perfectly at ease.

"How much more shall the blood of Christ, who through the eternal Spirit offered himself without spot to God, purge your conscience from dead works, to serve the living God?"

Paul had a good conscience before God, yet God held him in condemnation. Paul was a sinner even though he was working as hard as he could to please God. He was not guilty of the sins of modern Americans. Like any normal man, he suffered from concupiscence, but under the Law, there was forgiveness for sinners, and he offered the necessary sacrifices. His religion was not sufficient to save him. Just like the religion of the 3000 converts on the day of Pentecost, religion was a hindrance to his salvation. America is **cursed** more and more every day by the rise of cults and false religions of every sort. The New Age movement, Mormonism, and Islam are exploding in membership. These religions are a certain way to perdition for they will salve the conscience, and deceive the sinner who thinks he has found God.

Those Jews on the day of Pentecost who heard Peter preach cried out, "...*what shall we do?*" Yes, there is something the sinner must do. When Christ rose from the dead, our salvation was perfected forever, but we must receive it. First of all, there must be a pricking of the heart by the conscience. This pricking of the heart is the revelation to the sinner that he is a sinning sinner who has insulted his Creator and Redeemer. The sinner must confess this to God. God already knows it, but our confession is an agreement with God that we deserve to go to hell for our sin of disbelief. Next, we must call upon Him, and ask Him to save us on the authority of the atoning work of Jesus Christ, His own Son. There can be absolutely no plea of personal merit of any sort. If you hold back on iota of good works to claim for yourself, to offer to God for merit, God will not save you. You must come to the cross completely devoid of all you good works.

A GUILTY CONSCIENCE

The world calls it a guilt complex, and believes it is a sort of mental illness. Psychologists and psychiatrists take in a great deal of money for counseling these people, calling them "patients." A guilt complex can be a very serious threat to the health of an individual, because it can be very painful. The psychologist is interested only in making his client feel better. Right and wrong will not be considered in most diagnoses or treatments of mental illness. If another person has been wronged by the client, there is little likelihood that the psychologist will try to get his client to rectify that wrong.

The Bible is a better book of medicine than the ones used by the psychologist or psychiatrist. The Bible was written by the One Who created the human organism, and does give the best remedies for all sorts of illnesses, real or imagined. God created man in the image and likeness of Himself. Conscience means a moral awareness of right and wrong, and that is exactly what God wanted to protect man from - the knowledge of good and evil. God placed a tree in the midst of the garden of Eden as a symbol of His authority. God commanded Adam and Eve to leave the fruit of this tree alone. We know what happened. They disobeyed God and ate of the fruit of the tree. They were not punished for eating fruit. They were punished for disobeying God.

"And the eyes of them both were opened, and they knew they were naked; and they sewed fig leaves together, and themselves aprons." (Genesis 2:7).

The result was that they immediately knew right and wrong - they had a conscience. As a result of the sin of Adam and Eve, and the whole human race, which was in them at the time, the conscience is not to be trusted until it is energized by the Holy Spirit. A person living in God's world without God cannot follow his conscience because it will lead him astray. Eskimos steal their neighbor's food without any problem from their conscience. Sometimes the conscience will steer a sinner right, and sometimes it will not. It is not to be trusted.

A guilty conscience can be very painful, but it can be ignored until it becomes less effective as the heart hardens. A good conscience and a tender heart are part and parcel of the same happy person. When our conscience smites us, the time to take action to restore it, is right then. Listen to the prayer of David as he repents of his sin in response to the pain of his conscience.

Psalm 32: *"Blessed is he whose transgression is forgiven, whose sin is covered. Blessed is the man unto whom the LORD imputeth not iniquity, and in whose spirit there is no guile. When I kept silence, my bones waxed old through my roaring all the day long. For day and night thy hand was heavy upon me: my moisture is turned into the drought of summer. Selah. I acknowledged my sin unto thee, and mine iniquity have I not hid. I said, I will confess my transgressions unto the LORD; and thou forgavest the iniquity of my sin. Selah."*

We cannot know for which sin **David** is praying, but twice we are told that **David**'s heart smote him, and there are other sins in his life beside. We wonder how a man with so many serious sins in his life could be ". . . *a man after mine own heart. . .*" as God called **David** in Acts 13:22. The reason is that under the Law of Moses as well as after Calvary, man has always been forgiven and saved by grace following the sinner's repentance, confession, and faith. **David**'s repentance was not confession only, and certainly not confession to another man, but his confession followed the smiting of his heart which produced pure repentance. He was sorry after a godly manner.

"Now I rejoice, not that ye were made sorry, but that ye sorrowed to repentance: for ye were made sorry after a godly manner, that ye might receive damage by us in nothing." (II Corinthians 7:9).

Godly sorrow is a recognition of the true nature of the sin, and a determination never to do it again. We may fail, and do it again, because some sins are not so easily overcome, but we hate that and strive to overcome it. **David** was loved by the LORD because he

- 34 -

was a man who had a tender heart, and responded to the smiting of his heart by his conscience in the right way. Notice that **David** said in verse 2 that the man to whom the LORD imputeth not iniquity is a man in whom there is no guile. This man does not come to the altar pleading for grace with the intention of using grace as a salve for his conscience while he goes out to do the same thing all over again. Repentance is, among other things, a turning away from the sin that has aroused an angry conscience. The offspring of God hates his sins. His flesh may be ever so much in love with sin, but the new man hates it.

David suffered from his conscience while he kept silence, verse 3. His "... ***bones waxed old though my roaring all the day long.***" His weeping and wailing was affecting his health and taking away all the joys of life. There was no pleasure in anything even for the king of Israel. His life was as dry as a potsherd, though he lived in a palace with everything at his command. There was no let-up. His conscience beat him night and day. He could not sleep for his conscience did not need to sleep. "... ***night and day thy hand was heavy upon me ...***". verse 4

At last **David** relented and acknowledged his sin to God. He gave up trying to hide his sin. He called it iniquity. It was not just a pleasure he had enjoyed. If this sin were the sin he committed with Bathsheba, he could have called it a pleasure enjoyed by two consenting adults. That modern cliché won't impress a displeased God. Iniquity is deep and crimson sin. Adultery is a sin of grave consequences to those involved in it, and grievous suffering to those affected by it. There is no such thing as a sin that doesn't injure other people in some way.

In Psalm 38 we hear **David** crying

"***My wounds stink, and are corrupt because of my foolishness.***"

David's conscience had smitten him and wounded him. His wounds could be healed only by the divine forgiveness of a gracious God. Reading this psalm reveals that **David** was actually physically sick because of his conscience. The load of iniquity he bore was crushing him to death. In verse 18 of that psalm **David** says, "***For I will declare my iniquity; I will be sorry for my sin.***"

God knows of every good thing we do, and people do many good things, and He knows of every bad thing we do. There is nothing hidden from the all-seeing eye of the LORD God. Even the king's chamber was not off limits to God. The private business of the king was the judicial business of the supreme Judge of all the earth. The personal business of every person on Earth is under the divine scrutiny of the LORD God.

The remedy for an evil conscience is repentance, confession, and faith. A godly sorrow is inflicted upon us by our conscience. We must judge the sin and confess it - agree with God that it is truly sin. We must believe that Christ died to save sinners, and that He will abundantly pardon those who come to Him by faith. We can claim I John 1:9,

"If we confess our sins, he is faithful and just to forgive us our sins, and to cleanse us from all unrighteousness."

Only after God has forgiven and cleansed can our conscience return to a quiet and peaceable state. **David** Never had complete peace after his sins. He knew he was forgiven, and his conscience no longer smote him after his confession and forgiveness, but no man who has any integrity at all, can ever completely put out of his mind and heart the knowledge that he has greatly sinned and injured another person. I think that on the inside, that **David** went to his grave crying,

"... O my son Absalom, my son, my son Absalom! would God I had died for thee, O Absalom, my son, my son!"

THE CONSCIENCE

We continue our study on the conscience Thank God for a brain to think with; and a converted spirit to understand God's Word with; and the opportunity to study. Since the beginning of human history, man has sought knowledge, and especially knowledge about God. Our main problem as humans is, that we come to the study of God's Being and work with so many preconceived notions, that the truth has a great deal of difficulty penetrating our closed minds. And I say "I" because I am as prone to do that as anybody else. All sorts of hindrances come rushing in as soon as we take up the Bible and begin to strive for more understanding of what is written therein.

I have many friends in the pastorate. Among these pastors are some of the most devout and learned men in the ministry. When I am with them, I like to talk about the things of the Word of God, and delve into dark and hard sayings. The fundamentals of the faith which we cannot compromise are very few. Outside the fundamentals of the faith, I believe the brethren should have all the liberty the Holy Spirit allows them to have. I would not think of breaking fellowship with a friend over something like eschatology, or the theophanies. You may say that I am a liberal - I am. I am liberal in allowing other people to enjoy their God-given rights. Many pastors are afraid to discuss hard or unusual subjects, though, because of the fear of being ostracized. A man who has convictions that go outside the basic fundamentals of the faith becomes judgmental and unkind and ungracious. Not very good attributes for a pastor. I must confess that I am not immune to doing that myself on occasion.

I am afraid that this fear of being ostracized or being accused of being a heretic has stifled our study and writing. There is very little literature of any great spiritual value being published today. I believe that every pastor should be a writer, and with the easy access to and operation of computers, there is actually no excuse for a pastor to fail in producing literature. A pastor can write and produce tract booklets now that could have an enormous effect on his field of ministry. I don't know how much the 60 page booklets I produce costs, but I am sure that they would be much less than a dollar apiece. It is easy to make audio tapes, but an audio tape cannot always convey the rich spiritual content that is necessary for today because they cannot have the care in production as printed Words on a page, and they are not easily rewound to repeat a hard-to-understand sentence. Teaching audios are often bed time stories, rather than lively discussions. It is easier to defend one's position, too, if it is committed to Words on paper. It is also easier to be attacked. A fundamentalist should not be concerned about being labeled a heretic.

We must have more good readable literature that is basically sound and Scriptural because our understanding of the Word of God is deteriorating. We see this decline in the type of music that people want to hear, as well as their appearance when they come to church. The sermons of great preachers in fundamental period- icals is not a fully adequate substitute for books that delve into the why's and wherefore's of what that great preacher is saying.

I would love to see a church institute a program of required study for Sunday school teachers. I am talking about a doctrinal study led by the pastor, preferably written by the pastor, in which church history is studied in detail. Doctrines that are rarely heard of should be studied so that the Sunday school teacher will know why he is talking about what he is talking about. We have neglected the education of our young people to the extent that young people feel the church has no substance, and certainly no relevance to their own lives. Young people are interested in creation, but few there are who can satisfy their hunger for answers to serious questions. Most young people lose interest in church, and drop out because there just doesn't seem to be anybody in church who really knows what he is talking about.

We have been studying the conscience on this program for the past six broadcasts. We use the Word "conviction" in our Christian vernacular, but as I have already pointed out, that Word is not even in the Bible. "Conscience," on the other hand, is in the Bible some 30 times. A human cannot fully understand himself unless he understands his conscience, and what it does for him, or to him, or with him, or whatever. I will publish this series of broadcasts in a book, and you are welcome to receive one as soon as I can get it published by just asking.

There are several ways to define the conscience so that among these definitions there will be one that makes most sense to you. For example, we can say that the "Conscience is that faculty by which one distinguishes between the morally right and wrong, which urges one to do that which he recognizes to be right and restrains him from doing that which he recognizes to be wrong, which passes judgment on his acts and executes that judgment with his soul. Webster defines conscience as the sense or consciousness of right and wrong."[1]

Another has said that it is "... a consciousness of a court within man's being or the categorical imperative,"[2] and others have said that "conscience is the ethical sense organ in man."[3]

Conscience is innate. Read Romans 2:14,15 which tell us that conscience is within us and universal.

Every human has a conscience, at least at birth.

Conscience is in three parts. (1) Obligatory. It urges man to do that which he regards as right and restrains him from doing that which is wrong. (2) Judicial. Conscience passes judgment upon man's decisions and acts. (3) Executive. It condemns man's actions when his actions are contrary to what the conscience judges to be right.

To speak of an "erring conscience" is a misnomer. The conscience does not err, but the standard on which the conscience acts may be in error.

A morbid, perverted, or narrow conscience is one that is out of proper balance, narrow, fanatic, or bigoted.

A pathological or neurotic conscience originates in a psychic disorder or in a neurosis related to phobias, obsessions, fixed ideas, and compulsions.

A doubting conscience is one that acts in uncertainty. Romans 14:23 declares such actions to be sinful.

A dull, callused, or dead conscience is a condition wherein conscience ceases to function because of repeated disregard of its warning voice. Paul speaks of it as a *"seared conscience"* in I Timothy 4:2.

A good conscience is when man acts in conformity with his convictions. The Formula of Concord says, "Faith cannot exist and abide with and alongside of a wicked intention to sin and to act against conscience."

A social conscience is the merging of the individual moral consciousness into a group moral conscience.

Freedom of conscience is the freedom to believe, practice, and propagate any religion whatsoever or none at all.

Conscience is a guardian of morality and justice, and decency in the world.

The conscience is an irrefutable testimony to the existence of God. It is irrefutable evidence for the creative work of God. Conscience is more than just the knowledge of good and evil. Conscience is an active part of our being, and will smite us and punish us for violations of what it thinks are right. The conscience will convict us of wrong, and punish us with disquietude, distress, shame, and **remorse**. An accusing conscience will take away all the joy of life, and leave one hating himself.

1. "Elwell Evangelical Dictionary", other data unknown.
2. Ibid.
3. Ibid.

JOSEPH'S BROTHERS

The greatest success story ever told is the story of **Joseph** as recorded in the book of Genesis. **Joseph** was the eleventh son of his father and the first of two sons of his mother, Rachel. **Joseph** was a handsome lad, but perhaps thought a little higher of himself than he should, but he was born to be a servant of God, and God used even that to bring about His perfect will in the lives of this family. Because he was the first son of his father's favorite wife, his father doted on him probably a little too much, and the other boys saw that and resented it. God's will from the beginning was that one man and one woman join together for life as husband and wife, and the wisdom of that can be seen especially in our day when people take and dispose of spouses like they do automobiles, and children are torn between their real parents and the parents they have to live with. Few parents have the wisdom to make such a situation a success.

In **Jacob**'s family everybody worked. What a blessing is work. Work can cure a host of problems. The ten older boys were away from home tending the flocks when **Jacob** decided to send **Joseph** to see about them. He wore his fine coat of many colors, and when his half brothers saw him coming, they saw an opportunity to rid themselves of this pest. They sold him into slavery to a caravan of traders who took him to Egypt, and sold him to a very prestigious Egyptian soldier.

Potiphar's wife took a lust for **Joseph**, and failing to seduce him, she slandered him, and her husband threw him into prison. Evidently, the husband was not too sure that his wife was telling the truth because he did not execute **Joseph**. With the power of God upon him, and divine plans involving him, **Joseph** had special powers to interpret dreams. This power brought him before Pharaoh, and Pharaoh made him the second ruler in the kingdom because **Joseph** prophesied a famine was coming, and the king perceived that Joseph's God had His hand upon him.

After a couple of years of famine, which affected the land of Canaan where **Jacob** and his murderous sons lived, **Jacob** heard that there was food in Egypt, and he sent the ten older brothers to Egypt to buy food for them. **Joseph** recognized them, and plotted at once with himself to bring his whole family to Egypt because there were yet five years of famine to go. This family had to survive this famine because from this family would come the promised Deliverer. **Jacob** was number three in the line to Christ, and one of his sons must continue the line which must reach into the future another 2300 years to the birth of Jesus.

Joseph had no desire to harm his brothers though he certainly had the power to do with them whatever he wished. He remembered the bitter years he had spent as a slave and as a prisoner. No man could forget such experiences. But his God had been with him all the time, and now his heart was as tender as a child's and he wanted only to be reconciled to his family. His plans were successful and his family did arrive in Egypt where they stayed and increased for many years until the Exodus. He carefully led his brothers to the point he could forgive them.

We want to think about the emotions of **Joseph**'s brothers during all this time. In Genesis 37 where the tale begins, we find that when his brothers sold him as a slave, they were happy just to be rid of him. All except Reuben, the oldest son. When he found his little brother was not in the pit where they had thrown him, Reuben tore his clothes in anguish. However, in verse 27 we are told "... *And his brethren were content.*" These men were herdsmen, living much of their time in solitude. It is hard to imagine that their conscience did not smite them. They sat around their supper fire and filled the night with mirth.

Incredibly, the next day, they hatched out a plot to lie to their father about what had happened to his favorite son. They took Joseph's coat and soaked it in the blood of a lamb they killed for the purpose, and took it to their father, and told him that an animal had killed **Joseph**. Old **Jacob** almost died with grief, but there is not one iota of conscience among these ten sons. What they did to their old daddy was just as bad as what they did to their kid brother. Yet, they felt no **remorse** whatever.

Now these men stand before **Joseph**, the second ruler of Egypt, terrified for their lives for he had set a trap for them, and now they were squirming in it. In chapter 42 they talked among themselves in **Joseph**'s presence, not knowing he could understand them for he talked to them in Egyptian through an interpreter. In verse 21 they say,

"... We are verily guilty concerning our brother, in that we saw the anguish of his soul, when he besought us, and we would not hear; therefore is this distress come upon us."

"... and be sure your sins will find you out." says Numbers 23:32. The sins of these men had slept within their hearts like a rabid wolf, but now something had awakened the thing, and it was scaring them to death. Man can lull the monster within him to sleep, but you can be sure that sooner or later something will wake it up.

"... their hearts failed them, and they were afraid, saying one to another, What is this that God hath done unto us?"

When conscience awakens, it comes up with a fiery, sharp sWord. When your conscience finds you out, you won't feel so bold.

These men had to go back home and get their baby brother, Benjamin, the little son that cost Rachel her life, and take him to Egypt. After much persuasion, and seeing the hopelessness of the situation, **Jacob** allowed his precious little boy to go to Egypt, for **Joseph** had demanded that he be brought. No one in the family had any idea that the ruler of Egypt was their **Joseph**.

On their second visit to Egypt, **Joseph** brought them to his home, and again the men were terrified, and plead for their lives. They are now sure that they are going to be killed, and Benjamin with them. But they are allowed to start home again, but again **Joseph** tricks them, and brings them back to him. Judah pleads for their lives.

"And Judah said, What shall we say unto my LORD? Or how shall we clear ourselves? God hath found out the iniquity of thy servants: behold, we are my LORD's servants, both we, and he also with whom the cup is found."

The situation is now virtually hopeless, and the men despair of their lives. Judah not only confesses, he tells of their sin in detail.

True repentance brings a confession to God with no excuses and no alibis. A smiting conscience leaves us with no excuses.

Judah confesses that their sins will have further dire consequences. Sin just keeps on bearing bitter fruit. Like a seed of a thistle bringing forth thousands of thistles, sin bears a harvest of pain and despair. In verse 31, Judah says,

"It shall come to pass, when he seeth that the lad is not with us, that he will die. . .

"If **Joseph** keeps Benjamin in Egypt, his father will die,

"... and thy servants shall bring down the gray hairs of thy servant our father with sorrow to the grave."

Joseph at last told them who he was, but they could not believe it at first. They could not speak to him because *"... they were troubled at his presence."*

It is not until after old **Jacob** died in Egypt that the brothers finally begged **Joseph** for forgiveness. Their conscience had smitten them, and they had confessed their wrong, but it took the stark reality of their perilous situation to bring them to at last beg **Joseph** for forgiveness. Genesis 50:17,

"... and now, we pray thee, forgive the trespass of the servants of the God of thy father. And Joseph wept when they spake unto him."

Joseph wept tears of joy because finally, there was reconciliation. Now he could completely forgive. He said to them,

"Now therefore fear ye not: I will nourish you, and your little ones. And he comforted them, and spake kindly unto them."

Joseph is a type of Christ, showing the Savior's love and grace.

WORK OF THE CONSCIENCE

Romans 2:14 says, "*For when the Gentiles, which have not the Law, do by nature the things contained in the Law, these, having not the Law, are a Law unto themselves.*"

God gave man a conscience when he sinned and came to know good and evil in the garden of Eden. God gave the conscience to man for the preservation of the race. After man became aware of good and evil, his fallen nature would motivate him to love evil and hate good. Many of those who do good, do it to keep their conscience from troubling them. Doing good is not a sign of faith in God always, but is a response to an accusing conscience. Just prior to the flood, man seems to have completely lost his conscience, and violence filled the earth such that man was about to exterminate himself.

Look at the condition of mankind at this time. Read Genesis 6:5

Man had reduced himself by sin to the point that he was not able to have even an imagination of what good was. Like many people today, they cannot think of anything but evil. Even their imagination was evil. As a result there was no respect for human life, and violence was the business of the day. People were killing each other with a great slaughter. There must have been bodies everywhere piled up with no one to care enough even to bury them.

I believe that one of the reasons there are so few human **fossils** is because there was only a handful of people left by the time God gave the command to Noah to build the **Ark**. Man's conscience had completely deserted him.

According to Paul's Words, there are Gentiles, or sinners, who behave themselves in a righteous manner. There are, and always have been, sinners who choose to be kind, courteous, and all such things. Many people choose to be that way simply because it feels better than being the opposite. The environment can influence people to be good or bad. Economic conditions can help people to be good. It is good when people can get along together.

Paul says that such people do not have the Law of Moses, but they are a Law unto themselves. The show the work of the Law written in the hearts, and their conscience bears witness to them that they are doing well. God said in Romans 1:19,

"Because that which may be known of God is manifest in them: for God hath shewed it unto them."

We are talking about sinners. Sinners come in all categories and degrees. Good sinners need salvation just as much as the bad one needs salvation, and he needs the same salvation, and he must get it in the same way. Earthly good behavior will not redeem a man's soul. God gave man a conscience he could trust, but man has abused his conscience until he can no longer trust his conscience.

This goes back to what I said before about the conscience being influenced by **cultural** mores and standards. A good work ethic in a group will tend to make a better society, and people will be better. Where food is plentiful, stealing is not an accepted part of the society. The conscience will adjust itself to whatever a person must do to survive. Where survival is easier, the conscience will be more tender and right in its judgments.

The conscience bears witness of all that we do, according to verse 15. It is as if man has a judge living inside of him that watches everything he does, and comforts him if he does right, and smites him if he does wrong. The conscience seems to work entirely separate from our own mind, and we seem to have no control over it. It either accuses us or excuses us, and it is independent of our conscious thoughts.

There must be a distinction made between the conscience of a redeemed person and a sinner, and we will discuss that in a later broadcast.

Before the Law of Moses was given, all men had to depend upon their conscience. Let me give you some instances where Gentiles were a Law unto themselves. In Genesis 12:18,19, Abram has forsaken the place of blessing because of a famine in the land, and has gone off into Egypt. It seems that God's people were often prone to go off into Egypt during times of hardship, and Egypt becomes a type of the world. Because his wife was such a beautiful woman, Abram, for his name had not been changed yet, told the Egyptians that Sarai was his sister. Sarai wound up in Pharaoh's house, and Abram was well entreated because of her. But God plagued Pharaoh, and he somehow came to know that Sarai was Abram's wife. Pharaoh called Abram in and blessed him out for lying to them, and said,

"What is this that thou hast done unto me? why didst thou not tell me that she was thy wife? Why saidest thou, She is my sister? so I might have taken her to me to wife, take her, and go thy way."

Evidently, at this early time in the history of man, some men yet retained a righteous conscience. Conscience is not the ultimate standard of goodness, and it may often deceive us, but energized by the Holy Spirit, it is always reliable.

Abraham's son, Isaac, did the very same thing years later. Like her mother-in-Law, Rebecca was a beautiful woman, and like his daddy, Isaac was afraid the men of Gerar would kill him for his wife. Years passed, and one day Abimilech the king, saw Isaac and Rebecca sporting together. Abimilech called Isaac in and dressed him down for his deception. Listen to the king,

"What is this that thou hast done unto us? one of the people might lightly lien with thy wife, and thou shouldest have brought guiltiness upon us."

Even this heathen Philistine displayed a righteous conscience.

In Genesis 39:9 **Joseph** revealed the sensitivity of his conscience when his master's wife tried to seduce him. *"... thou art his wife: how then can I do this great wickedness, and sin against God?"* An act of adultery is not just an innocent time of pleasure between two innocent adults; it is wickedness, and a sin against God and each other.

Nehemiah would not take benefits from the people even though the former governors had done that, and had even allowed their servants to rule over the people. The former governors had robbed the people, but Nehemiah would not do so. He said, *"... but so would not I, because of the fear of God."* Nehemiah had a sensitive conscience, and it was motivated by the fear of God. All of

- 47 -

us, from the cradle to the grave need a good healthy fear of God in us. One of the charges brought against me by the parents in Cherokee County, Alabama when they were demanding that I be fired for teaching their children the Bible was that I taught them the fear of God. They said their churches did not teach that God was to be feared. Pitiful churches that do not teach the fear of God. It is the fear of God that makes survival on this planet possible.

Psalm 19:9 says, "*The fear of the LORD is clean, enduring for ever: the judgments of the LORD are true and righteous altogether.*"

Nehemiah had that. Nehemiah's fear of God enabled him to behave with integrity toward the people he was leading. Daniel demonstrated what a true conscience is when he refused to eat the meat from the table of the great kind of Babylon, for he

"*... purposed in his heart that he would not defile himself with the portion of the king's meat, nor with the wine which he drank; ... .*"

A good conscience will take away carnal pleasures. A good conscience is like a sleeping dog – best left alone. That is why Paul commands us to "*Abstain from all appearance of evil.*"

EVOLUTIONIST'S CONSCIENCE

I wrote this essay on the 1st day of November, '96, in Deland, Florida, and I hope to be home for Thanksgiving. That is the place to be on Thanksgiving and Christmas. Wonder why that is. There is so little regard for home and family among Americans today, compared to what it used to be, but we still have that warm and homey feeling on those two blessed holidays. Both of those holidays are a part of our American heritage - our Christian American heritage.

I Timothy 4:2 says,

"Speaking lies in hypocrisy; having their conscience seared with a hot iron."

People can speak lies in hypocrisy after they have destroyed their conscience. A living thing that has been seared has become hard and unfeeling. I met a man once at a rescue mission who had been burned over most of his body, and he was angry and bitter because of his condition. Most of his skin was as hard as dried leather, and without sensitivity. He could not feel the touch of another person. A conscience that has been seared still exists, but it is encased in unfeeling, uncaring emotions. A seared conscience will no longer smite the individual, and then he is free from its restraint. Tragically, many people want to be free of conscience.0

An illustration of how it is possible for a person to speak lies in hypocrisy is seen in the Anniston, Alabama museum. This museum is not unique in its dishonesty and deception, however, for virtually every museum has displays in it that are deceptive. The Anniston Museum has a painting of a fossil bird named Archaeopteryx. There have been five of these **fossils** found in Europe, and a few fossil feathers. Casts and other types of reproductions have been made of the best of these fossils, and this one appears in many places. Pictures of it are found in many science books, including textbooks used by school children.

The bird is a weird looking creature, having teeth and cLaws on its wings, and when it was discovered, evolutionists were wild with joy because they had at last found a transitional fossil. They claimed that this creature was a transition between reptile and bird. It was a creature that was part reptile and part bird. This rejoicing among evolutionists lasted until they at last had to admit to themselves that this was not a part-reptile, part-bird as they claimed - it was nothing more than an extinct bird. Today evolutionists admit among themselves that Archaeopteryx is just a bird. What a person believes has more influence on his world view than what he knows.

Yet the painting in the Anniston Museum has a caption under it claiming that this is a transitional form of a reptile-bird. They know this isn't true, but there it is nevertheless, for the tender eyes of boys and girls to see and believe.

I do not believe I am being radical to assert that this is no less than a crime against humanity. Any act that is destructive of human happiness and well being is a crime to some extent against humanity. No good thing can come from deceiving people. Telling lies in hypocrisy cannot bring forth any lasting benefit to the person who is lied to. I have a very strong aversion to calling people liars, and even using the Word "lie," but it is the only way to describe what is being done here.

This museum is controlled by a committee of local business people who are all leading citizens of the community. I am sure that many of them are members of local churches. All of them, I am sure have good motives, and spend a good deal of their hard earned money supporting this museum. Some of them, perhaps all, would vehemently deny that they are evolutionists, and they may not be. They are supporting the evolutionist lie, nevertheless.

I wrote a letter to the director of this museum, pointing out the error in the museum. My complaint fell on deaf ears - not even answered by the director. I gave full documentation for what I said in the letter, which was as kind as such a letter could be. Doubtless, I was shrugged off as some fanatic who was not worth listening to. The lie goes on. There is a museum in Tampa, Florida, the exact name of which I cannot recall, but it is a very fine museum, costing the city and its citizens a lot of money. Great numbers of school children go in there every year. It is established as a learning center. People who go there expect to gain valuable knowledge. They have confidence in such a place, and they should be entitled to believe what they learn in any museum. Unfortunately, that is not the case.

This museum has done an outstanding job of obtaining a cast of a dinosaur track way which originally was found in the Paluxy River in Texas. They have mounted this track way on a wall in a hallway so that it can be closely examined by the public. It is a very fascinating, and no doubt popular exhibit. But there is a very serious problem with this display. At the end of the cast there is a plaque with information on it telling how the dinosaurs became extinct millions of years ago.

In the river bed of the Paluxy River there are prints in the stone with the dinosaur tracks that are almost certainly human footprints. These prints have been subjected to scientific study by qualified scientists, and the findings have shown that the prints were indeed made by something stepping there, and pressing down the soft mud back when that stone was just mud. I have a picture of a footprint showing the three lines across it where it was sawed apart so that cross sections of it can be examined. I have no problem believing that it is a human footprint, though I cannot say that it is because of the uncertainty of some of the scientists I know who are reluctant to say that it definitely is.

The truth is that there is absolutely no way of knowing how old the dinosaur footprints are. No scientist can say those dinosaur footprints are 65 million years old because he has no way of knowing. I have discussed radio metric dating several times on this broadcast, and pointed out why it is not reliable as a method of dating field samples of rock. Statements about the age of fossils are pure speculation.

There is good evidence that human footprints are found in the same strata along with the dino footprints. And here is my charge: that plaque is a lie in hypocrisy because if tells something for a fact that is not a fact, and leaves out something that may be true. That is propaganda, indoctrination, half truth. It is completely dishonest, and without excuse. It is teaching a religious dogma in violation of the Constitution of the United States, and it is an outrage.

Why should we get so upset about this? Why should we apply a verse of Scripture to this sort of thing when the verse has a more primary application? Why should we get upset when people are being told something that is not true? The most rabid **evolutionists** in the world ought to get upset when people are being told something that is not true when that hurts people. But they don't get upset because their conscience is seared with a hot iron. Their conscience has been branded with Satan's brand.

A pastor friend and his father in law and I were eating at a restaurant when we struck up a conversation with the waitress. She said she believed in evolution. I asked her why she believed in evolution and she said, "Dinosaurs." I asked what dinosaurs had to do with it, and she just shrugged her shoulders. The conversation went on for some time, and ended with her saying she had heard the gospel many times, but she was comfortable with what she believed.

This young woman no doubt will experience shipwreck in her life. What will be her fate? Will her conscience remain so dead that it cannot be awakened to her soul's deep need? So much hangs on the efficient function of the conscience.

Evolution does not free anybody from anything. It is a bondage no less restrictive than any other false religion. Having a conscience seared with a hot iron may free one do live comfortably with what he believes, but it does not free him from his responsibility to his Creator. When we face God, the only ones who will be comfortable will be those who have hidden themselves in the grace of our LORD Jesus Christ.

THE NATIONAL CONSCIENCE

In Psalm 9:17 God tells us very clearly that *"The wicked shall be turned into hell, and all the nations that forget God."*

Occasionally, I have the bad experience of meeting a wicked person. A truly wicked person is one who will lecture you about how the Bible is foolishness; how that there is no God; or God is just some notion in people's mind. The ungodly know it all. They have all the answers. Every suggestion by a believer is snapped off by a demand for proof. I met such a one. He was 82 years old. Raised in an orphanage, he had been to Sunday school all his childhood, and judged it all a waste of valuable time. Expressions of concern about him as a human being with a soul that will never die are countered with sneers that man has no soul and death is the end of everything. This man had no a conscience.

Nations, too, can have consciences. A conscience is that part of a person which guides him to do what he believes is right, and smites him when he does that which he believes is wrong. America once had a tender conscience. America was founded on the principles of the Word of God, and so America had a conscience that could be trusted. For the most part, the people who governed America were people of integrity though we have surely had our share of scandals through the years America knew what was right and what was wrong and pretty well behaved accordingly. Our Laws and mores reflected the nation's conscience. At one time in our nation, the sale and consumption of alcoholic beverages was illegal. Women wore long dresses, and men wore clothes that revealed no more than a sense of self respect. Children could go out to play and be relatively safe. Rape and incest were not so common. Now all these crimes are broadcast as news for public entertainment.

When I was a boy the religious beliefs of people who believed what our forefathers believed were tolerated. Now, we are scorned and our rights hardly recognized. We are considered enemies of human rights. There were very few cults, and people either went to church or they didn't, but seemed like everyone had respect for the church. Every religion in the world can be taught in the state schools in America, except the true Christian faith upon which the nation was built. Christianity has been out lawed. At last it became the Law of the nation that school children could no longer be exposed to the Bible in their schools, or hear prayers offered to their Creator and Redeemer.

One of the greatest indicators of the state of a nation's conscience is its culture. In fact, a nation's conscience and its culture are inseparable, and, in truth, may be the same thing. If we understand better what culture is, then we can have a more positive influence on it. Culture has to do with the tilling of the soil so that it will be cleaner and more productive. Soil that is clean of undesirable plants and insects will bear more and better produce.

Culture has to do with not only the arts but also every aspect of the nation's life. The work ethic of citizens is affected by the state of the culture of the nation. The educational system was affected by the state of the culture. Religions of every sort are affected by the state of the culture of the nation. You and I are affected by this state, or condition, or level of culture in which we live. Even people who take no interest in culture, and who feel it is something for other people, are affected by the culture in which they live.

Culture has to do with what we might call the finer things in life. Fine gold is gold that is free from impurities. A coin that is classified as fine in a coin catalog is one that is almost totally free from blemishes. A fine person is one that all of us desire to be around, and have as a friend. A fine person is one who has the fruits of the Spirit:

". . .*love, joy, peace, long-suffering, gentleness, goodness, faith, Meekness, temperance*. . ." (Galatians 5:22,23).

The **Supreme Court** should be the conscience of the nation. It is to the Supreme Court that people appeal when they believe that their Constitutional rights have been abused. Being a judge is a very heavy task. Sitting in judgment of other people is not a matter for the unwise or unbalanced. Truth and justice must reign supreme in a court of Law. Justice can only be carried out where truth is honored. When a court passes the death sentence upon millions of innocent children, the conscience of the nation is hardened and corrupted.

A columnist wrote that the recent reelection of President Clinton was like the O.J. Simpson trial. It is this man's contention that Americans are so well satisfied with a seemingly healthy economy that they were willing to turn a blind eye to the moral issues involved. "Americans," he wrote, "disregarded the truth and decided for a not-guilty political verdict. The question, of course, is why?"[1]

The reason why is the theme of my essay. America has lost her national conscience. Mammon has taken the place of God in the hearts of the majority of Americans. Cal Thomas writes that Republicans "... did not raise many of the other **cultural**-moral issues about which government can do something."[2] Compare the thought of a man sitting in the White House who has a reputation like Mr. Clinton, with a man sitting in the White House with the reputation of a man like Abraham Lincoln, or even Harry Truman, and the picture becomes clearer. We have said to every boy and girl in America that it is all right to be an adulterer. It is all right to kill a baby as long as its head has not seen the light of day. It is okay to smoke dope as long as you don't inhale.

A nation with a defiled conscience is on the brink of disaster. Just like a person with a defiled conscience, a nation will self-destruct. America should clothe herself in sack cloth and ashes on inauguration day, and fly black flags from every staff. We can expect a stock market collapse that will shake every person in the land. The president can be expected to pardon every criminal involved in the numerous scandals that have thrown shadows over him, including his wife, and perhaps he himself will wind up eventually being pardoned by President Albert Gore.

Our essay jumps forward from ten years ago to the present. Today Barack Obama is the president. He is a man that stepped out of the shadows, and swept the nation off its feet. His personal history is unknown and suspicious. His election is suspicious. He has appointed people to his administration that compose a gang of murderers, tax evaders, and other like characters. Every citizen who retains the old-timey conscience of our forefathers mourn the present state of our nation. It is depressing to realize that more than half of our citizens fall on the side of radical liberals who have brought our nation to the brink of failure. We will soon see in operation those great Words:

"The wicked shall be turned into hell, [and] all the nations that forget God." (Psalm 9:17).

Hear these ominous Words:

"So [are] the paths of all that forget God; and the hypocrite's hope shall perish:" (Job 8:13).

But then, you can't expect a man who has sat under the preaching of a rabble-rousing racist for twenty years to know anything about the Word of the living God. It is not Mr. Obama who upsets me so much as the foolish majority in this nation that elected him. People without a conscience could not be expected to vote for a wounded old soldier when they had a good-looking hunk at hand; especially when that hunk has a smooth, though crooked tongue.

This preacher is not the only person in this nation who mourns November 5, 1996. Many of us remember the sweet, gentle America where there was a national sense of integrity that demanded that our leaders show at least a little of it. We mourn the disappearance of our national conscience. We hate evil because it hurts people. We hate evil even because it hurts evil people.

---

1. Thomas V. DiBacco, *Voters' Decision Like O.J. Simpson Decision:* "The Orlando Sentinel", 11-7-96, p. A

2. Cal Thomas, *Four Unpleasant Years Ahead:* "The Orlando Sentinel", 11-7-96, p. A-23

WORD STUDY

John 8:9 says,
"And they which heard it, being convicted by their own conscience went out one by one.".
When the sins of these hypocrites were brought out into the light of reason and justice, their conscience sent them scurrying. What is this powerful force in the human being?

It is relatively easy to understand man's tri-unity. Man is composed of body, soul, and spirit, and the Bible teaches us that in passages such as I Thessalonians 5:23 which says,

"And the very God of peace sanctify you wholly; and I pray God your whole spirit and soul and body be preserved blameless unto the coming of our LORD Jesus Christ."

We were created in the image and likeness of God, and God is a triune Being. We know man had a conscience in both the Old Testament and the New Testament because even though the Word "conscience" is not in the Old Testament, we see it at work in numerous occassions such as I Samuel 24:5 where it is said that **David**'s heart smote him.

Whatever the conscience is, and however man got it, or when, evolution cannot explain man's having it. What we are told in the Bible is all we have to go on. It is to our benefit to understand our conscience because the more we know about ourselves, the more effective we can be as Christians, and the more effective our soul-winning efforts will be. The best way to understand our conscience is to study what the Word of God says about it. I like to think of the Bible as an "owner's manual." As the Maker of mankind, God has provided a complete manual giving instructions on the maintenance and operation of our whole being.   As a Bible literalist, I seldom need to fall back on Word studies to ferret out the meaning of most passages of Scripture I speak on. In the case of the conscience, I have been benefited by studying the Greek and Hebrew Words that apply to the Word "conscience." Man was created by God, and is therefore a highly complex organism. Man is more than just flesh and bone, even though most people behave as if that is all they are. The physical body of man is complex enough, leaving physicians wondering often how this or that could possibly work. Medical science has indeed advanced to a very high state, yet it is actually quite primitive when we consider how much physicians do not know.

The most important two thirds of man is virtually unknown at all, and as far as I know, no **evolutionist** has put forward a theory, make that myth, to explain the life of man which we know as the soul and spirit. Most evolutionists just deny that such things exist. Involved with the soul and spirit is the conscience. Shipwrecks are disastrous, but a compass helps avoid them.

The Bible uses the Word "heart" extensively to speak of man's inner being. The Word "leb" and its derivatives, have variously been defined as "ravish," "become intelligent," "heart," "understanding," and "bread." "Concrete meanings of leb referred to the internal organ and to related physical locations. However, in its abstract meanings, 'heart' became the richest biblical term for the totality of man's inner or immaterial nature. In biblical literature it is the most frequently used term for man's immaterial personality functions as well as the most inclusive term for their sin, in the Bible, virtually every immaterial function of man is attributed to the 'heart'."[1]

The Apostle John uses the Word "conscience" once in his Gospel, but he does not use the Word in his epistles. Instead he says,

"For if our heart condemn us, God is greater than our heart, and knoweth all things."

What John is saying sounds like the passage in I Samuel 34:5 where we are told that "... *David's heart smote him, ..."* Our conscience and our heart are very close to being the same thing.

Conscience has to do with the Word "log - is - mos' ", which means computation. The conscience has to do with our ability to reason. It involves our imagination and thought. Computation is figuring out, and the conscience "figures out" what is right and what is wrong for us. It is the computer that sends out alarm signals when we go astray.

Conscience also has to do with the Word "as - then - eh' - o" which means to be feeble, or diseased, sick, or weak. When we go astray, our spiritual life is not well. Our spiritual life can be sick to a greater or lesser degree, but whatever the cause is, we know it, and we must take immediate steps to remedy the situation. If we have a bruised arm, it may get better by itself, or it may get worse and need the attention of a physician. In any case, care should be taken to give the bruise opportunity to heal, and if it doesn't, then a doctor is needed. "As - then - ay' - ma" means infirmity, a weakness caused by neglecting the conscience.

In discussing the wicked behavior of the Corinthians around the LORD's table, Paul says,

"For this cause many are weak and sickly among you, and many sleep." In the next verse, Paul says, *"For if we would judge ourselves, we should not be judged."*

The Corinthians had allowed their sin about the LORD's table to defile their conscience, and as a result some were sick and some had even died. People must pay attention to their conscience, and judge themselves according or there will be sore trials.

Another Word, "kat - ag - in- o' - sko" is related to conscience is which means to know something against, or to know by experience, and therefore to think ill of, or condemn. In Galatians 2:11, it says,

"But when Peter was come to Antioch, I withstood him to the face, because he was to be blamed."

Paul acted as Peter's conscience. Peter's conscience was enlightened and exercised by being condemned in the sight of others. Peter could have thrown a temper tantrum like is common among God's people today, when he was condemned for his inconsistency and sin, but he had a heart that brought him under self-condemnation and repentance. His conscience smote him.

When our conscience knows something against us, and tells us about it, we should go to the LORD with whatever is necessary to make our life right again, and to satisfy our conscience. It is foolish to live in misery with an accusing conscience when it so easy to take our trespasses to the LORD.

Our next Word is "soon - I' - day - sis." This Word means "a knowing with," or a co-knowledge (with oneself). When Paul withstood Peter to the face, Peter already knew he was to be blamed, and therefore, he was not able to rightfully respond to Paul with alibis. He could have covered up, as many are prone to do, but he wanted to get the matter right so that his life would be productive, and his heart could be at peace. When an obstinate heart stands against conscience, then the person is unhappy and fruitless. He will not be able to walk with the LORD in peace, and he will soon disappear from the ranks of the church.

At last we come to "toop - to" which means to thump, cudgel, or pummel. This Word means a continual hitting as with a stick. The conscience can beat a person until he is at the point of death. This Word means to beat, smite, strike, wound, and our conscience can do all of those with a vengeance. The remedy for whatever is causing the conscience to smite us may be a bitter pill to swallow, but it is better to swallow the pill than to be beaten by the conscience. Pride is the hardest nut for the conscience to crack. People will stand in church during an invitation while their conscience thrusts a sWord through their heart, but they will not move because they have pride that is like an impenetrable wall.

David's heart smote him, and he repented. Peter was rebuked publicly by another apostle, and he repented. What if they had not? Would **David** have written the beautiful psalms? Would Peter have written two great epistles found in the New Testament? Does God have a great task for you to do, but is blocked because you have sinned.

---

1. _Theological Wordbook of the Old Testament_

NORTHERN KENTUCKY

The conscience is a valuable part of man's makeup, and it can be a good friend, and we wonder how some people can behave in unconscionable ways. How can a person do things that are wrong and cruel, and never seem to be troubled by his conscience?

In Boone County, Kentucky at present there is a virtual war being waged by humanists and other religious bigots against a creation ministry that wants to build a creation museum in Boone County. You would think that in a society seemingly concerned about education, fairness, and justice, the citizens of any given community would welcome a museum. Not so in Boone County. The Answers in Genesis ministry, led by Mr. Ken Ham, bought land and asked the zoning committee to recommend to the full commission that the land be rezoned for a museum. Fortunately, the zoning committee was not composed of religious bigots. They made an objective decision, and recommended to the full commission that the land be rezoned.

Mr. Ken Ham, Executive Director of AIG, writes that a local minister said, "We wouldn't mind if they wanted to build a church but they want to build a creation museum to change our minds." Perhaps this minister would not be so happy if somebody wanted to build a church that would change people's minds.

This situation has caused an uproar in northern Kentucky, with the media giving a great deal of time and space to it. Mr. Ham writes, "The 'Free Inquiry' group, together with other humanists, have led a very active campaign to try to stop the county from granting us permission to build this museum. (So much for the name 'Free Inquiry!') They contacted newspapers and television stations to make this a big media event."

According to Mr. Ham, however, the media couldn't get its teeth into the issue because they couldn't figure out why a nonprofit organization, with private funding on privately owned land, couldn't build a museum with a Christian perspective. That is indeed a good question, but we must remember that this is America, 1996 - the America that just reelected Mr. Bill Clinton as president - the nation with the defiled conscience.

The humanists and their allies care nothing about rezoning land - they are opposed to Christianity because Christianity is not just religion, but the faith founded by the Creator Redeemer.

One lady called Mr. Ham a "Jim Jones," but Mr. Ham is not Jim Jones, because Jim Jones was receiving all sorts of government checks from his slaves, and had full government sanction for everything he did. Mr. Ham wants nothing from government but just and equal treatment under the Law.

Of course, the separation of church and state raised its phantom head, as opponents threatened lawsuits if the county allowed the museum to be built. A minister from a local United Church of Christ joined the opposition saying that he believed in evolution. In a meeting in a local church, the opposition gathered with a lawyer to devise some conspiracy by which they could prevent the museum from being built. A lawyer at that meeting said, "If on the other hand, we lose, we're going to be in very bad trouble. A meeting like this will be illegal. You'd go to jail. You might get killed. Now, as it's happened in the past. Don't think for a minute it can't happen now. It's happened throughout human history. What do you think Bosnia is over? That's what's going on over there."

That sort of rhetoric is intended to build a hysteria that is completely unjustified. Bible-believing Christians - and it is sad that we must say "Bible-believing" Christians, because the Word "Christian," like the Word "gay," has been derogated and slandered until its meaning has changed to something else - Bible-believing Christians have never been a threat to anyone. Christians will defend themselves when they have their backs to the wall, but no true Christian has ever been a physical or political threat to anyone. Religious liberty did not spring out of humanism. Religious liberty, the basis of all liberties, sprang out of the Word of God, for; after all, it is the goodness of God that provides good things for His creatures.

A local minister named Rev. Adams, according to a local newspaper, said," '... he doesn't want the county's children to be taught that God intends them to be bound to ancient Scripture'." Mr. Ham continues in his article: "This minister sent us a letter listing affirmations his church holds to, including: 'We further urge pastors and teachers to teach about the problems of biblical literalism'."[1]

Those leading the opposition to this museum are evidently radicals who will go to any length to force their values upon not only northern Kentucky, but our whole nation. America was not founded by or for people who hold such hatreds in their hearts that they lose all sense of fairness and justice. Even the Gospel of Jesus Christ is considered hate language by some.

The human conscience could not tolerate such violently immoral behavior if it were not defiled. Paul wrote to Titus,

"Unto the pure all things are pure: but unto them that are defiled and unbelieving is nothing pure; but even their mind and conscience is defiled."

A defiled conscience is one that is still living, not like the seared conscience which is dead, but the defiled conscience awakens for the wrong reasons, and will lie dormant when it should be smiting. To attack something as innocuous as a creation museum sponsored by a relatively small and harmless institution is the work of reprobate minds with conscience so confused they cannot consider truth as a possibility.

Unbelievers are defiled, whether they be ordained by some religious institution; whether they be members of legal institutions; or whether they be members of ecclesiastical orders, or anything else. Without a working conscience that is tuned to the heart of God, man can be a monster. The opposers of the museum are dangerously near being monsters. One man threatened to place satanic emblems on his property which adjoins the museum property.

Christians will tolerate just about anything in the interest of peace. Christians are a working people who ask only to be left alone so that they can rest for the day's work. Christians do not rush into the streets to riot and burn and kill every time they have some real or imagined grievance. Most of us take little action to aid our brothers and sisters when they are attacked. Our conscience should be more tender.

In I Timothy 4:2 a seared conscience is spoken of. *"Speaking lies in hypocrisy; having their conscience seared with a hot iron."*

That verse speaks of religious zealots who are far from God and His grace, and are under the direct influence of Satan's emissaries. Having a conscience branded with a hot iron reminds us of old cowboy movies showing cowboys branding steers out on the old prairie. When a steer is branded, that shows who he belongs to. When a human has a seared conscience that shows who he belongs to. A person with a conscience that approves things that Satan approves, and hates the righteousness of God, has been branded by the devil himself. God speaks of those *"Who hate the good, and love the evil,"* in Micah 3:2. Such people, God says,

"... pluck off their skin from off them, and their flesh from off their bones; Who also eat the flesh of my people, and flay their skin from off them; ... ."

Many people who have protested the abortion trade have experienced virtually such treatment. Women have had lace ripped from their clothing, and hair bows and ribbons torn from their hair. Arms and legs have been twisted and secured by vicious restraining devices. People have been dragged about like so many animals.

An unbeliever is dangerous enough to himself and those around him, but when an unbeliever has a defiled or seared conscience besides, then he is a real and present threat to those he disagrees with.

---

1. Ken Ham, "From Boone to Bosnia!" <u>Answers in Genesis!</u> (October, '96): 1

THE TESTIMONY OF COSCIENCE

Paul wrote in I Corinthians 1:12,
*"For our rejoicing is this, the testimony of our
conscience, that in simplicity and godly sincerity,
not with fleshly wisdom, but by the grace of God,
we have had our conversation in the world, and
more abundantly to you-ward."*

Because he had a good conscience, **Paul** was able to rejoice
because his conscience was clear. A troubled conscience is a good
way to get sick. A clear conscience is maintained by obedience to
the commands and concepts of the Word of God. I fear that in
preaching salvation by grace that we give the impression that good
works have no place in the life a saved person. That is not true.
When we are saved,

*"... we are his workmanship, created in Christ Jesus unto
good works, which God hath before ordained that we should walk
in them."*

A clear conscience is tied to these good works. A
disobedient child is an unhappy child. Note the puffy lips, the
redness of the cheeks, and the rage in the eyes of a rebellious child.
This is why a child should be made to obey. It is less painful for a
child to receive a good paddling than to be allowed to remain in a
state of rebellion. The conscience of the child is smiting him, and he
will suffer consternation and frustration if he is allowed to continue
in his disobedient state.

A child of God is no less unhappy when he is rebellious. Of
course, over the years, a rebellious heart will defile the conscience,
and it will give up it efforts to bring the wayward one back into the
will of God. He will never feel entirely good about his life, though,
and at the end he will have regrets and **remorse** over his wasted life.

A good conscience is a necessary element in the physical and mental health of any person. Paul was so happy and well adjusted in his clear conscience that he could rejoice even in his persecutions. Paul's conscience testified "... *that in simplicity and godly sincerity* ..." he had had his conversation in the world, and particularly in the church at Corinth. The Christian speaks to the world with his behavior even more loudly than with his Words. Our testimony to the world must be good if our conscience is to be good. Our testimony to the world is the good works to which we were created when we were born again. (Ephesians 2:10).

II Corinthians 13:5 teaches us that there are two reason for obedience, or submission: an external one and an internal one. In the case of that particular verse, the external reason for obedience is the sWord, and the internal one the conscience. Generally speaking, the external one is the wrath of God, and the internal one is the conscience. The wrath of God can be anything from physical illness to the contempt of the world around us.

Evolution cannot explain the conscience, any more than it can explain any other metaphysical entity of man. Yet the spiritual aspects of man are far more important and complex than man's physical body. Evolution attempts to explain man's behavior by studying animals. Animals do eat, reproduce, and interact with each other, but assuming that man is an animal is the wrong basis for studying animals. The spiritual parts of man were created by the same Creator Who created his body, and God's Word is the only source of sound information about the complexities of man's spiritual being.

There can never be a good conscience toward man until there is a good conscience toward God. Man cannot be wrong with man, and right with God. Man cannot understand man's values until he has held communion with God. Man must be explained by God. The psyche of man is not a subject for laboratory experiment. Studying man as an animal gives false results, and leads man astray in his understanding of man.

The conscience can only give a true testimony when it has been purged, or thoroughly cleaned.

"How much more shall the blood of Christ, who through the eternal Spirit offered himself without spot to God, purge your

- 66 -

conscience from dead works to serve the living God?" says
Hebrews 9:14.

Dead works are the problem, not the works of an obedient
servant. Dead works are the efforts of a sinner to save himself. It is
the precious blood of Christ that cleans the conscience of all the bad
influences of society and culture, and brings it into conformity with
God's will. "The conscience must adjust itself to the meridian of
eternity before it can tell the self what time it is, what duty is, and
how duty is to be done."[1]

Read Hebrews 9:19.

An evil conscience has been polluted with the culture we
live in, or by our own mistaken notions of the truth. An **evolutionist**
with an evil can attack God and think he is doing a great service for
his fellow man. No matter how intelligent a person may be, he is
subject to being a fool in God's sight if he has developed an evil
conscience, and denies the God Who created him. In that verse we
learn that the conscience and the heart are not exactly the same thing,
but that the conscience acts upon the heart.

This verse hearkens back to ". . .The Levitical ceremonies
with reference to the preparation of the priests for their priestly
service."[2] The priests had to undergo this ceremony every time they
ministered before the LORD. This typifies our approach to God.
We can "...*draw near with a true heart in full assurance of fait... .*"
when our heart is sprinkled with the blood of Jesus Christ, purging,
purifying, our evil conscience. We cannot come before God with
assurance and joyful confidence if we have unconfessed sin in our
life, and our conscience is so evil that it is ignoring that fact, or if our
conscience is smiting us, and we are ignoring it. I John 1:9 promises
us that we can have that cleansing if we only confess and ask for
forgiveness for our sins.

Redeeming help comes only from God. The inventions of
man, no matter how costly or elaborate or impressive, cannot touch
the garment spotted with the flesh. The pit of depravity is much,
much too deep for the puny engines of man. A skilled physician can
pry into the flesh of man, and do marvelous things, but only the
Great Physician Who shed His blood for the redemption of man's
soul can prod into man's spirit. Having a psychologist lead us to
confuse our conscience so that it becomes evil and confused is no
satisfactory cure for a troubled conscience. The soul and spirit are

forever, and the remedy for their ailments must be eternal. We should be glad that God has provided all we need to have a good conscience that can bear a good testimony for us.

All of man's problems are sin problems. It is not only with the skin that we have to do, but with the inner man, which is not seen, but is felt. The simplistic notions of man will not provide lasting relief. Talking to an unknown person behind a lattice work will relieve the pain of a weak conscience, but it will not suffice to heal the good conscience.

"Thou hast made us in a fearful and wonderful way: how complex is man! what a terror to himself! sometimes what a joy! now burning with intolerable agony, and now as it were on the wings of eagles, away up where the light is born, and heaven is fully seen. How abject, how August is man! Help us to study ourselves in the light of thy revelation, in the light of thy daily providence; enable us to ask great questions, to put reverent but fearless inquiries; may we not stand back in superstition and wondering ignorance, but approach quietly, lovingly, hopefully, to ascertain what we may of the mystery of things, and be ennobled by a higher veneration, softened and chastened by a sweeter consciousness of thy presence."[3]

———
1. Joseph Parker, *Preaching Through the Bible* (Grand Rapids, MI: Baker Book House, 1987), p. 269.
2. Kenneth Wuest, *Word Studies in the Greek New Testament* 4 vols. (Grand Rapids, MI: Wm. B. Eerdmans Publishing Company, 1989), 2:181
3. Ker, *Preaching Through the Bible*, p. 263.

MY BROTHER'S CONSCIENCE

Cain asked the impudent question, "... *Am I my brother's keeper?*" (Genesis 4:9b). The rest of the Bible reveals God's answer to this question. We are positively our brother's keeper. But who is our brother? Is our brother the same as our neighbor, whom Jesus spoke of at length? Our brother and our neighbor must be the same, for when Cain asked his question, all men were his brothers. The Word "brother" is used pretty loosely in the Scriptures, so we must interpret it as the context requires.

In I Corinthians 8:11 the brother spoken of is the brother in Christ. This is the most common use of the Word in the Epistles. We are altogether responsible for our brothers and sisters in Christ. At the beginning of church history, most Christians were converted idolaters. They worshiped idols with animal sacrifices and feasts in the idols' temples. After their conversion, they ceased this practice, but sometimes were invited by their unsaved neighbors to join in these feasts in a social way. A question arose about whether or not they should engage in such practices.

Most knew, as Paul affirms in verse 4 that an idol is nothing, and so eating meat offered to these things was no sin, and neither was it a sin for them to eat in an idol's temple. Christian liberty allowed them to do this as long as nothing was involved but their own personal consciences. Some did not know this.

Christians, like all other men, do not live unto themselves, nor die unto themselves, and what we do is of extreme importance to those around us. No Christian is at liberty to do anything outside the commandments of the New Testament, that would harm somebody else in any way, whether he is saint or sinner.

If a weak brother thinks that eating meat offered to an idol is a sin, then to him it is a sin. In fact, Paul says in verse 11 that the weak brother is apt to perish. If he sees a stronger Christian, or a Christian with more knowledge of grace, eating meat in an idol's temple, becomes a stumbling block to the weak brother. The weak brother who feels it is a sin to eat meat in an idol's temple would be emboldened to eat meat in the idol's temple in violation of his conscience, and that would be a sin to him. It is amazing to me that Christianity has survived so long in view of the carelessness with which Christians live the Christian life.

It seems to me as if this entire portion of I Corinthians has been completely ignored by the church. Chapters 8 through 10 tell us clearly that grace gives us the liberty to live as we please where the Bible does not place restraints upon us, but we do not have the liberty to violate the conscience of others, nor behave in such a way that our actions would cause other people to do things they think are wrong just because they saw us do it. Paul makes this point in verse 12 where he writes,

"But when ye sin so against the brethren, and wound their weak conscience, ye sin against Christ."

Some people think it is okay to go to movies when there is a film that seems innocuous enough, but there is always the possibility of wounding the conscience of a brother who feels that movies are of the devil, and because of the example of the other Christian, he goes to the movies in spite of his conscience, and sins. I would be one of those who feel that movies are the work of the devil, but I wouldn't go just because I saw anybody go into one. Not everybody is like me. Some would feel if they saw some great preacher go into a movie that it would be okay to go, even when his conscience still told him it was wrong. When we violate our Holy Spirit energized conscience, we will sin every time.

I am convinced that it is wrong for Christians to buy and sell on Sunday. There was a time when all Christians had this conscience concerning the LORD's Day. When the world destroyed the blue Laws for materialism's sake, the church lost her conscience about it. I did not lose mine, but I will not judge others.

Dr. John Rice wrote, "The first time God ever mentioned the Sabbath to men or gave commandments about it is in Exodus 16. The heavenly **manna** was given to Israel for six days and they were told that the seventh day was the Sabbath and they were not to gather on the day (Exodus 16:23-30). "God first made known the holy Sabbath to the Jews when they were gathered in the wilderness near Mount Sinai.

"But one may insist, 'The Sabbath command is in the Ten Commandments.' Yes, it is, but, remember, the Ten Commandments are addressed primarily to Israel.

'And God spake all these Words, saying, I am the LORD Thy God, which have brought thee out of the land of Egypt, out of the house of bondage. Thou shalt have no other gods before me'.

That is addressed to Israel."[1]   Dr.Rice points out that the New Testament outlines our liberty when Paul writes in Colossians 2:16,17,

"Let no man therefore judge you in meat, or in drink, or in respect of an holyday, or of the new moon, or of the sabbath days; Which are a shadow of things to come: but the body is of Christ."

People saved by grace need no superstitious rituals or ceremonies or holydays, for all of that pertains to idolatry and paganism. But Christian liberty is not as wide as some may think.

The truth is that our liberty is as narrow as our brother's conscience. The truth is that our liberty is as narrow as what the world perceives as Christian behavior. When I was working my way through college, I collected insurance for a living. I had to go into a bar to collect one premium. I was not altogether comfortable in going into that place, but I was there on business. If a sinner who knew me saw me go in there, it could have cost me whatever testimony I had, and that person could have gone to hell because of that. If another Christian saw me go in there, he might have figured that he could then get by with going in to have a drink, and he would have sinned by doing so.

I take the attitude Paul took. Listen to him in chapter 9 as he reviews his position and credentials. Paul has great liberties which he has not taken advantage of because of his concern for the conscience of others. In verse 19 he says,

"For though I be free from all men, yet have I made myself servant unto all, that I might gain the more."

Paul was more concerned for the souls of men than he was of anything else. I believe that is what gets a man close to the heart of God.

"The fruit of the righteous is a tree of life; and he that winneth souls is wise." (Proverbs 11:30).

I do my pitiful best to avoid allowing others to judge me in

"Wherefore, if meat make my brother to offend, I will eat no flesh while the world standeth, lest I make my brother to offend."

God forbid that one should stand among the damned at the judgment seat of Christ and accuse me of doing something that caused them to reject the atoning blood of Jesus Christ. I would hate for some waitress to point at me, and tell Christ that she would have gone to church if she did not have to serve me on the LORD's Day.

I am aware of the attitude of most Christians today about the LORD's Day. I know that I may be chalked up as a legalist. I know all of those things. I will judge no man in regard to anything pertaining to his liberty in Christ, but to my dying day, by God's grace, I will prod people to consider their ways. God has given us His mind when He said, *"Abstain from all appearance of evil."* (I Thessalonians 5:22).

------

1. John R. Rice, *In the Beginning* (Murfreesboro, Tennessee: SWord of the Lord Publishers, 1975), p. 99

Introduction to Book II

A miracle is a miracle because it is something only God, or a person He empowers, can do. Miracles were a tool God used to validate the claims of Jesus that He was God's Son, God incarnate. Miracles still take place, but they are very rare, and not for public view. They are not done to prove the claims of the Bible. Since the Bible was finished and canonized, human faith had to be placed in the Bible, not works. A miracle requires power over matter which only the omnipotent God can exercise. Matter has Laws governing its behavior. Certain of these Laws man has discovered; there may be other Laws that man has not yet found. Gravity is a Law. Decay is a Law. Motion is a Law or series of Laws. These Laws govern matter even in outer space, because they govern matter, and matter is found in all areas of the cosmos.

Man has learned much about matter, and if he would apply what he knows about matter and its Laws, he would not be clinging to the myth of evolution. Accepting the hypothesis of evolution causes a scientist to become a philosopher. He has ceased to be a scientist because he is willing to accept unproven and unproveable data as evidence. Indeed, he has accepted it as a Law. It is not possible to believe contradictory ideas, and remain in the tight circle of rationality.

Objects brought from the field should always by looked at suspiciously. Finding a bone fragment sticking out of the earth is interesting, but to prove that that bone is millions of years old is another matter. The bone cannot be considered as a laboratory specimen because it has been exposed to the weather for who knows how many years?

The business of a scientist is to use his five senses to examine things in an objective way, having no preconceived notions beforehand, except the hypothesis he is working on. He must give an honest and documented report of his findings to the public. Anything that cannot be examined with a microscope, and other instruments cannot be judged accurately. Evolution philosophers make rediculous statements, and they get away with it because we live in a world where scientists like this hold the power of public opinion. People want to believe evolutjon because it eliminates God. People are free to do their own thing in the absence of Jehovah. . A

- 73 -

scientist works with objects he knows the history of, and he keeps accurate notes. Nothing can be true or reliable unless another scientist can duplicate the experiment, and get the same results. Scientists develop things that are useful to man. Men like Isaac Asimov, Stephen Hawkins, Stephen Jay Gould, and many others, spend their time dismissing God as if He were a human weed.

Man creates things: paintings, buildings, poems, etc., but his creating work involves things that only the Creator can create. The man-creator must use created molecules of paint, molecules of steel, molecules paper. Man uses created matter to create everuting he creates. Even the brain that makes such wonderful and beautiful creations is an object off the Creator's drawing board. What an amazing creation is the body of the man-creator.

Nearly all of Jesus' miracles could fall into this book, for few were the miracles that did not envolve creation.of new matter.

Book II

CREATIVE MIRACLES

The first miracle of our LORD is recorded in John chapter 2 and verses 7 through 11. The miracle occurred at the marriage feast at Cana, early in the ministry of our LORD. I am sure scholars know a great deal about the customs of that day, and they could explain a great deal to us about the drinking of wine, and just exactly what it was. Dr. John R. Rice asserts that it was not fermented wine, but somehow I am not convinced that it was not. Customs do not determine what truth is, but customs can make an enormous difference in how people do things or see things. Let's read the passage, and hear what the LORD has to say. I urge you to read this passage before you go on.

The water the servants poured into the water pots was pure water. They knew it was pure water, because they had drawn it, and poured it there so the guests could wash their hands after the manner of the Jews. We learn in Mark 7 how that the Jews had found fault with Jesus' disciples because they didn't wash their hands before they ate. Beginning in verse 3 we are told,

"For the Pharisees, and all the Jews, except they wash their hands oft, eat not, holding the tradition of the elders. And when they come from the market, except they wash, they eat not. And many other things there be, which they have received to hold, as the washing of cups, and pots, brasen vessels, and of tables."

Folks, these Jews were as fastidious as could be on the outside. They were too nice to be nice.

In John 3:25 we hear of a clash between some of John's disciples and the Jews about the matter of purifying. We are given no more information about that clash, but we can assume that the disciples got tired of being nit-picked by the Jews over the hand-washing bit, and they had a few choice Words to say about it. The hand-washing was a tradition of the Jews, and was not a part of the Law. There was a laver at the entrance of the tabernacle, and it was there for the purpose of the priests washing themselves before they entered into the service of the LORD, but the Jews had added their own rules to the Law of God, and had run it into the ground. Who can criticize anybody for washing their hands regularly, unless the hand-washing has something to do with making the person fit to serve God? Little boys, including this one, have a wholesome animosity about washing too much. I strive to keep my spiritual garments clean and white, but my hands - well, they don't need so much attention.

If this water had borne any real significance, the Savior would probably not have changed it. But it was water for a religious purpose without any real significance, like so many religious things man does or says, that have no real spiritual meaning or value. How comical it is to see a man dressed up in gorgeous garb, swinging an elaborately designed and decorated thing with smoke coming out of it, making people think that polluting the air can benefit the human race. Jesus said to the woman at the well, *"Ye worship ye know not what:..."*. He could have added, "And ye worship ye know not how." People ought to wash their hands before they eat always, but for sanitary purposes, not pretense of being spiritually clean.

After the servants had filled up the great big pots, the Master simply told them to draw out some of the liquid, and take it to the governor of the feast. Somewhere and sometime during those few seconds, Jesus performed a notable miracle. (Is there any such thing as a miracle that is not notable?) Water is hydrogen and oxygen - two gases. Both gases are invisible. Water can be separated into the two gases very easily. Jesus had created both hydrogen and oxygen when He created the heaven and the earth as recorded in Genesis 1:1. He knew what the water was. This was many centuries before man would learn that water is a combination of oxygen and hydrogen molecules.

Water is necessary for physical life. It takes about 800,000 gallons of water to grow an acre of cotton. The human body requires an intake of about 2½ quarts of water per day, a quart of it gained from drinking liquids, and the rest from the food we take in. This water is given off as sewage either through our pores or is passed off as liquid or in solid waste.

Water is composed of molecules having 2 hydrogen atoms and 1 oxygen atom. When Jesus turned the water into wine, it was necessary for Him to come up with 6 carbon atoms, 10 more hydrogen atoms, and 5 more oxygen atoms to add to each water molecule to make wine; for one type of wine is $C_6H_{12}O_6$. Of course, Jesus could have called these atoms from somewhere in the universe far away from Earth, but if He had done that, He might have upset something else. It is true that Christ knows at all times where every atom is and what it is doing. That is what Colossians 1:17 and Hebrews 1:3 tell us. Hebrews says,

"Who being the brightness of his glory, and the express image of his person, and upholding all things by the Word of his power ... ."

Christ upholds all things, meaning the atoms of the universe, as well as everything else. Christ did not need to call the atoms required for the wine from someplace else; He created them. This is a creative miracle. Most of our LORD's miracles were creative miracles. If Christ had created grape juice, as Dr. John R. Rice insists, He would have needed different numbers of different atoms, and He would have created them.

While in Russia I learned that Russian Baptists use real wine in their communion services. American Baptists feel this is unwise for good reason. Converted alcoholics would be placed at risk by having to drink alcohol in church. People who have never tasted wine may be tempted to drink it as a result of tasting it in church. If wine is served in church, then it gives the appearance that it is not as dangerous as the pastor has declared it to be.

In Russian communions, they fill a large goblet with wine, and pass it around the congregation. Each member takes a sip (or two or three?), and passes it to the next person. In some churches a deacon takes the cup each time it reaches the end of the row,, and wipes the brim with a towel before the next person drinks. In others, they do not do that. One of those I visited in Siberia did not wipe. (Talk about purification - give us some.). When the cup is empty, it is refilled and started around again.

The second chapter of John ends with a marvelous revelation. Verses 24 and 25:

"But Jesus did not commit himself unto them, because he knew all men, And needed not that any should testify of man: for he knew what was in man."

Jesus Christ's knowledge of humans is a very valuable asset to people. He knows how many hairs are on our heads, requiring Him to keep a running total. Each of us has His undivided attention every moment of every day. It is as if there were no other people as far as His attention to us is concerned. No, that does not mean that each one of us has a separate god, it means that God is so unlimited in knowledge and power that He can interact with every human alive on Earth and in Heaven as if there were no one else alive. This means that I am not interrupting somebody else's prayers when I begin to pray, for He is able to hear me and multitudes of others at the same time. If we have several little children gathered around us all chattering at the same time, we may be able to understand what two or three of them are saying. Jesus Christ understands all the little children around Him chattering at the same time.

The creative miracles of Jesus reinforce John 1:1-14 that introduce Jesus as God incarnate. The Creator was able to speak the universe into being in a breath, and this is shown to us in His creative miracles as they were done instantaneously. To compromise with evolution and believe that God needed billions of years to develop the universe and life in it, is a gross violation of Scripture, what we know about God, science, and common sense.

CREATED MANNA

Christ not only created the carbon atoms, and the additional hydrogen and oxygen atoms to turn the water in the water pots to wine, as we discussed in the last essay, but He must also have used a great deal of power to fuse those elements together into the wine molecules. The atoms had to be bound together in the right pattern in order for it to be wine. A miracle, according to the Kennedy definition, is the reversal, or suspension, of natural Laws, or the Laws of physics. The dictionary is a little more simple. The dictionary in my computer tells me that a miracle is something that causes wonder or astonishment. Many things cause wonder and astonishment other than miracles, and so I prefer my own definition. Further, according to my definition, a miracle can be performed only by the LORD God, or one to whom He assigns such power and authority.

Jesus reminded the Jews in John 6:49 that twenty-five hundred years earlier, the LORD performed a miracle for the children of Israel in the wilderness of Sin. They had been out of Egypt only 2 months and 15 days when it became evident that there was not enough food in the camp for three million people. Impending starvation is no small threat, but faith in God is no little power. Believing Jews that they were the chosen people of God, chosen for the purpose of producing the One Who would bless all families of the earth, according to God's promise to their father Abraham, as recorded in Genesis 12:3. The "mixed multitude" had seen numerous miracles performed by their God in Egypt. They had even walked across the Red Sea on dry ground, and seen Pharaoh's army drowned by the same sea as it pursued them to recapture them.

Why couldn't this people believe God? If God allowed them to starve to death, He would defeat His own purpose, and show

Himself in the eyes of the world incapable of caring for His own. His name would be destroyed. Joshua feared for God's (7:9) after the defeat of the Jews at Ai:

"For the Canaanites and all the inhabitants of the land shall hear of it, and shall environ us round, and cut off our name from the earth: and <u>what wilt thou do unto thy great name?</u>"

True faith in God does not rest upon miracles. Faith comes by hearing (Romans 10:17). I fear for people who claim that their faith rests upon some miracle they believe they have seen. Here is the reason: the children of Israel had seen many miracles, yet they had little, if any, faith in God. Faith does not come by seeing miracles, but rather by the hearing and reading of the Word of God, If you've never seen a miracle. don't worry about it. Leave the miracles for people who need entertainment.

People who have a legitimate complaint will go to the person, or persons, who are responsible for the problem, and ask for relief before they do anything else, These Israelites did not have a legitimate complaint. They were hungry in a wilderness. of course. Their children were hungry and crying, I know. But they did not have a legitimate complaint because they had a God Who was aware of their need, and would supply it at the appropriate time, if they would but ask and believe. They preferred complaining to asking.

The best way to bring the appropriate time to the present is to pray. Our circumstances are known to God at all times, and He cares about us, and He will respond with the solutions for our needs when we pray. Somebody said that God only moves in answer to prayer. God is sovereign, but he has given humans a free will, and He will not override that free will unless He judges it is necessary. Usually, We have not because we ask not. We have not because we ask amiss, lusting for something we don't need. I have to confess that I have asked amiss. There were times I didn't know for sure how to ask.

The children of Israel would have had a legitimate complaint if they had prayed, and God had not answered. God always responds to prayer in some way - though often not in the way we think or choose. If the children of Israel had been people of faith, they would have been on their knees appealing to God, and not whining to Moses and Aaron. Still, in Exodus 16:4 we read,

"Then said the LORD unto Moses, Behold, I will rain bread from heaven for you: and the people shall go out and gather a

certain rate every day, that I may prove them, whether they will walk in my Law, or no."

God would provide.

The manna that God provided was specially created for this purpose. Probably, there has never been anything like it on the earth before or since. We are not certain about what it was. The Apostle John called it bread in 6:31,

"Our fathers did eat manna in the desert; as it is written, He gave them bread from heaven to eat."

But in his sermon, recorded in Deuteronomy 8:3, Moses gives us a most interesting insight into what it was. Listen,

"And he humbled thee, and suffered thee to hunger, and fed thee with manna, which thou knewest not, neither did thy fathers know, that he might make thee know that man doth not live by bread only, but by every Word that proceedeth out of the mouth of the LORD doth man live."

Here we learn several things about this created bread. God allowed them to hunger. God humbled them. God made them know that man doesn't live just by bread, but by the Words that fall from the lips of God. A mature saint is one who has learned that the Word of God is food no less important than the food on the table. The nourishment of the soul can be ignored, but the consequences of ignoring it is a starved soul.

God created the **manna**, for there was nothing in the wilderness to bring it forth. This wilderness was a desert, with some scrubby bushes here and there, but there was no water on the surface. We know not what it was, and so we cannot guess what atoms went in to its making. We do know that it was a small round thing as Exodus 16:14 tells us, as small as a flake of frost upon the ground. They called it "**manna**" because they did not know what it was, and neither did their fathers know. It would not keep because the next day whatever was left over stank and bred worms, except on Friday, when they could gather enough to last them over the Sabbath. It was like coriander seed; it was white, and tasted like a wafer made with honey. Numbers 11:8 tells us that it tasted like fresh oil.

The Jews

"... ground it in mills, or beat it in a mortar, and baked it in pans, and made cakes of it: ...". Manna was very small, the size of a barley corn, and white. It must have been very potent in food value,

and contained everything a human being needed for complete nutrition because the Jews lived on it for 40 long years. They ate it until they entered the promised land. Joshua 5:12 tells us,

"And the manna ceased on the morrow after they had eaten of the old corn of the land; neither had the children of Israel manna anymore; but they did eat of the fruit of the land of Canaan that year."

There is an intriguing statement in **Job** that I only mention in passing, because as far as I know there is no other reference in the Bible to this. Psalm 74:14 says,

"Thou breakest the heads of leviathan in pieces, and gavest him to be meat to the people inhabiting the wilderness." I am not even sure if *"the people inhabiting the wilderness. . ."*

refers to the wilderness wanderings of the children of Israel, but it is sure that God fed somebody with dinosaur meat at some place and time.

Manna is a type of the Word of God, for the same God Who created **manna** created the Bible. The same God Who saw the wanderings of the children of Israel, sees our wanderings here below. We must learn to live by every Word of God, and to consider the Bible an indispensable part of our diet. As **manna** alone was sufficient for the Jews, the Word of God alone is sufficient for us. The Jews ate **manna** until they crossed over into the promised land, and there they ate of the old corn of the land.

Murmur not against your Creator and Redeemer, my friend. Do not complain of the hardships of walking in righteousness in a world of unrighteousness. Be faithful. Be unmoveable.

CREATED BREAD

A miracle is a thing of great interest, especially for people who observe it. We use the Word "miracle" quite commonly in our language, but actual miracles happen very seldom. Most of us have seen the exhibitions on TV of men performing what they claim to be miracles of healing, but most of us realize that there is something involved there other than miracles. People who had the power to heal the sick, had the power to raise the dead. No one would seriously claim to raise the dead today, but it might come to that.

While I was in Belarus a few years ago, I stayed in the home of a Russian family who were charismatic. I sort of got led into this unawares, but it was a good experience for me. The family invited me to their church, and so to be courteous, I agreed to attend one of the meetings. I do not remember the name of the evangelist who was an American, but, in my opinion, he was a liar and a shyster. He spoke for an hour to a large assembly of Russians who were eager to hear the Gospel. All they heard was the bragging of this man about the healing he had performed in other places. My hostess was a physician who was suffering from cancer caused by the Chernobyl explosion, and she was very sick. During the "healing" time, the evangelist urged people to place their hand on their sick part. Then he urged the people to come forward for healing. It was a complete disaster. A few people came, and he made a big deal out of them, but it was as phony as a three dollar bill, and the audience knew it.

I sat on the platform facing my very sick hostess. I watched her closely all during the healing time, and she never moved. She did none of the things the evangelist told the audience to do. On the way home, the family let it be known in no uncertain terms how disappointed and unhappy they were with this display of foolishness by this "great" healing evangelist. I said nothing. They were upset enough. They certainly seemed to have the same opinion of this man that I did, though I did not voice my opinion to them.

God does perform miracles today, but not for carnivals. God can create today, and probably He does, though I cannot give an instance that I know to be a creative act. Jesus performed many creative acts during His earthly ministry, and every one of them was for a purpose beside the benefit of the one healed. He was establishing His claim that He was the Messiah.

One of the most notable miracles Jesus performed was when he took five little loaves of barley bread and two little fishes, and fed a multitude that numbered upward of 15,000 .The record in the Gospels is that there were 5,000 men, and so there must have been at least that many women and children, and probably a great number more. This was a task of momentous proportions. The work of getting the food from the hands of Jesus to the hands of the people was staggering. The disciples surely passed baskets of food to designated men in the crowd.

The twelve disciples had just returned from an extended preaching tour, and after they had reported to their Master, they got into a ship and crossed the sea of Tiberias, which is the sea of Galilee, to a place near Bethsaida. This was open country with only a few villagers living around. Jesus was met there by a great number of people, and He and His disciples went up into a mountain, followed by this great host of people. This was in the early spring, but the grass had begun to grow, and though the Word calls it a "desert place," it was not a desert as we think of, but rather a place that was sparsely inhabited, and no surface water, and no trees. There on the mountain He was preaching and healing when they saw coming up the mountain another great number of people, and the crowd swelled to many thousands.

Of course, Jesus knew what He would do, but He wanted to hear what Philip would have to say about it. Jesus sent the disciples out into the crowd to see what they could find, and a little boy was discovered who had a small lunch, and this was secured for feeding this army of hungry mouths. Jesus always gave thanks for the food He was about to eat or serve, and this time was no exception. He then began to hand the loaves and fishes to the disciples, and the disciples distributed it to the groups of fifties and hundreds that were sitting around on the grassy slopes of the mountain. First, the loaves and then the little fishes.

How could this handful of food be divided to feed such a multitude? The truth is that the five loaves and two fishes were simply symbols. Jesus could have fed that multitude with nothing in His hands. His hands are creative hands. Creative in the true sense that He can create matter from nothing. The power of His Words is sufficient to bring forth matter, and the power of His hands could form that matter into the molecules of fish flesh and bread that fed that multitude. Jesus could have called atoms from someplace else in the universe for His purpose, because He is the guardian of every atom, but there are no atoms that are not filling a useful place in the system of the universe. It was better for Him to simply create the loaves and fish as He went.

How rapidly the Master had to work. This occurred as the day drew to a close, and so His hands had to move very rapidly to take a loaf, and break it in two pieces of the same size as the original, and hand it to a disciple. Perhaps the piece He created was much larger than the original piece He held in His hand. I may be a little foolish in my visualizing this, but I believe that the loaves Jesus created were the finest bread those people had ever eaten, and the fish probably had not a bone in them. I always think of our LORD as doing good things. The disciples wanted to send the people away, but the Savior wanted to bless them.

> Christ was the Word Who spake it,
> He took the bread and brake it;
> And what the Word did make it,
> *That* I believe and take it.

How foolish for man to try to explain this miracle away by teaching that Jesus used the little boy's lunch as an example of sharing, and motivated everyone in the multitude to share their food with those who had none. How could one explain the twelve baskets of fragments that were collected afterward if such were true? This was nothing but a notable miracle by the Creator Who has all power, and is able to do such things with little effort.

Dr. Henry Morris, in his new annotated Bible, in a note on page 1029, says, "This was nothing less than a mighty miracle of creation. Setting aside His own created Law of mass conservation (no matter can be either created or annihilated, as implied by Genesis 1:31 - 2:3 and set forth in the Laws of **thermodynamics**). Jesus, by supernatural power created a great amount of bread and meat to feed the multitude. This was well within His ability, as Creator of all things in the beginning (John 1:1-3; Colossians 1:16). This is the third of seven great creative miracles recorded by John, and it is one of the few events recorded by all four of the Gospel writers. Though there be some variation in the record of each of these witnesses, the details are exactly the same in each one."

The Creator created me unto good works (Ephesians 2:10). He created Adam unto good works, but Adam decided to go in another direction. I was not created by my good works, but unto good works. Pure religion is to visit the widows and orphans, according to James, and I fear I do not practice religion very faithfully. It seems that the Great Commission is on the top of my list of priorities. To preach the Gospel, the power of God unto salvation of lost souls, is what God has laid on my heart most heavily. The Creator Who created fish and bread, and indeed the whole cosmos, is able altogether to enable me to do what He demands of me. It is true for you, dear friend. If you are saved God has created you unto good works, and He has a work for you to do. Faithfulness is one of the most important elements in good works.

The world hates the teaching of Christ as Creator even more than it hates the teaching that Christ is Redeemer. By indoctrinating the public in evolution, sinners are reinforced in their unbelief, and saved people are made less devoted to the Saviour because they compromise His creative work by mixing evolution into it.

CREATED HUMAN TISSUE

Thermodynamics is a Word that seems difficult at first, and for a scientist it is such an important Word that books have been written about it. **Thermodynamics** is something that is important to Christians, too,, because **thermodynamics** is another of the means of regulating the universe that God established. The Word "thermodynamics" simply means "heat movement." "Thermo" meaning <u>heat</u>, and "dynamics" meaning <u>movement</u> are put together to name this Law of physics. All of us should know what a thermos bottle is, and what it does, and why. A thermos bottle keeps stuff hot or cold. A thermos bottle is made of two bottles, one inside the other, with a vacuum in between. The vacuum between the two bottles slows the movement of heat in or out of the stuff in the inner bottle, because heat does not move very well in a vacuum.

Christ <u>spoke</u> all matter into existence. This is recorded in Genesis 1:1 and Psalm 33:6. The power of His Word was sufficient to command stuff that didn't exist, and it obeyed Him. Therefore, Christ also maintains the inventory of matter in the universe. He knows where every atom is, and what it is doing. This is given to us in Hebrews 1:3, which should be read at this time.

And so Christ upholds all things "*. . .by the Word of His power. . .*".

Why should man think anything is impossible with God? This is a strange phenomenon. There is such a store of knowledge today, and the possibility of so much exposure to the Word of God, that it is incredible that there are people who still cannot accept the truth of God's omnipotence. The wicked heart of man that lusts to be equal with God is the hindrance to his ability to have the joy of knowing that God is supreme. The joy of being submissive to God is an element of a stable life and mind that only the believer in Jesus Christ can know   A rational person who knows God can accept the truth that God knows everything.   To believe that Christ knows where every atom is, and what it is doing, may be childlike faith, but it is childlike faith that pleases God. At the same time, that childlike faith can be mature faith that will do when one is dying.

There are three kinds of love: 1. Credo Deum="I believe that God is." 2 Credo Deo= "I believe God is true." 3 Credo in Deum= "I believe in God." i.e. I put my trust in Him. Number three brings salvation.

The truth that Christ is maintaining the inventory of atoms in the universe is consistent with the first Law of **thermodynamics** which states that no matter in the universe is being lost, and none is being created.   Miracles are miracles because they require a suspension of this Law, or some other Law, while the Savior performs creative miracles.

When a living thing dies, its body begins to lose atoms of matter immediately. Atoms of oxygen and hydrogen fly off into the air. To raise a person from the dead requires the **creation** of atoms to replace those that have gone off into the **atmosphere** to do some other task. Of course, Christ could have called those atoms back, and if someone chooses to believe that, it's okay. It would be a problem if someone chose to believe that Peter raised her by his own power. When Peter raised Tabitha from the dead, she had been dead for perhaps several days when he raised her as recorded in Acts 9:36 - 43.

Tabitha, whom they called Dorcas, lived in Joppa. She was a devout saint, loving the LORD and doing many good works, such as making clothes for other people, especially widows. She became ill and died, and her friends bathed her body and prepared her for burial, and placed her body in an upstairs room. The disciples sent two men to fetch Peter who was nearby in the city of Lydda. When Peter came, he took charge, and sent everybody out of the room. He kneeled down and prayed, and then looked at the body, and said,

"Tabitha, arise. And she opened her eyes: and when she saw Peter, she sat up."

As a result of this miracle, many people believed Peter's preaching, and were converted; and Tabitha enjoyed many more years of life on Earth.

The raising of Lazarus is one of the most popular passages of Scripture in the Bible. It is recorded only in John 11. Jesus was very close to Lazarus and his two sisters. They a short distance from Jerusalem in a little town called Bethany. Lazarus took sick and his sisters sent a frantic message to Jesus in Jerusalem to come quickly, but Jesus tarried for two days before going to Bethany, and Lazarus died. Jesus had a purpose in this, and though some may see it as cruel that Jesus waited until Lazarus died before He went to see about him, the LORD does things in His own way, and man had better not be a judge of God.

When Jesus drew near to the village, the sisters were mourning the death of their brother, and many neighbors had come in to lend their voices to the crying. Martha made haste to go out and meet Jesus, but Mary awaited His arrival at home. When Martha met Jesus, she cried out that if He had been there her brother wouldn't have died. She also believed that the LORD could raise Lazarus from the dead. It was to Martha on this sa LORD occasion that Jesus made that great statement,

"I am the resurrection, and the life: he that believeth in me, though he were dead, yet shall he live: And whosoever liveth and believeth in me shall never die. Believeth thou this?" (John 11:25, 26).

Martha then left Jesus and went home and told Mary that Jesus was waiting for her, and she got up and went out to meet Him. The mourners in the house thought Mary was going to the grave to weep there, and they followed her. When they met Jesus, and the Saviour heard their crying, and heard Mary's plaintive cry, He broke down and wept. Perhaps He wept because of the suffering He had called on them to endure because of what He wanted to demonstrate. Christ is aware of the suffering of His people on His behalf. Every missionary who ever died on a foreign field was watched over, and taken, by the LORD Jesus. Every martyr was an object of our LORD's mourning care as they died in His service.

All the grieving Jews were of the opinion that if Jesus had been with Lazarus, He could have prevented the man's death. Jesus wanted them to know more than that. He wanted them to understand exactly what He said to Martha, as recorded above. The notion that living for Christ is all ice cream and cherries is silly. God is engaged in a warfare with the prince and power of the air for the souls of men. This warfare is vicious and brutal, and God's people are in the middle of it. Soldiers get hurt - they often get killed.

When they arrived at the tomb, Jesus commanded the people to remove the stone that sealed the entrance to the cave. Martha objected that her brother's body by this time would be in such a bad state of decay that it would be creating an awful odor. That odor was the elements that had evaporated from the tissue of the dead man. His body was rapidly losing atoms of matter that Christ would have to replace in the exact order and position in order for the man to live again. Internally, his organs would be falling into a chaos of decomposition that would require great power and knowledge to restore. Only the Creator could raise this man from the dead.

The LORD Jesus demands that humans do all they can do before He will intervene and do that which we cannot do. Man can move a stone, but man cannot restore a corpse. When the stone was out of the way, not so the power of God could get in, but so the resurrected one could get out, Jesus called Lazarus' name and commanded him to come out of the tomb.

Of course, Lazarus' soul and spirit were in paradise, and that is where he heard the Savior's voice. Perhaps all the souls in paradise, including Abraham and all the old saints, heard Jesus call Lazarus' name. At any rate, Jesus called Lazarus' name so that all the dead knew that Jesus was calling forth only one dead one. Perhaps those in hell heard the voice of Jesus. There may have been a great outcry among the damned as they heard and recognized the Master's voice, but they were excluded from the call to life

A CREATED COIN

Money is good stuff. It represents our contribution to the society in which we live. In a very real sense, money is sacred stuff because it represents hours of our life which we have sold to others to make it possible for us to avoid dependency on our neighbors. Money can be sacred in a sense because it makes it possible for us to keep our bodies healthy. It pays for the food and shelter of our little ones who helplessly depend upon us. Money makes government possible, and human beings need government. Money makes it possible for us to maintain a suitable place to gather to worship our Creator, and send the good news to other peoples.

Money is evil stuff. When stolen by one method or another, and used to purchase stuff that damages our bodies, it is evil stuff. When money is gained by extortion or deceit it is evil stuff. Money can be used to hire evil deeds done. Money can be used to oppose the work of the Creator on this earth, and pervert the Gospel of our LORD and Saviour. Money is a power that can hardly be stopped.

"For the love of money is the root of all evil: which while some coveted after, they have erred from the faith, and pierced themselves through with many sorrows." (I Timothy 6:10).

Actually, money, as you know, can be neither good nor evil, but is only a submissive servant, doing whatever its owner chooses for it to do. Money is probably most evil when ministers love it

The LORD Jesus always fulfilled every obligation He had while He walked in this vale of tears. He had no paying job. It may be that some folks won't love Him because He was not on a big salary. He was not a rich and famous person. While he drew no pay, He earned His way. He had no stocks or bonds; no pension. He often slept on the hillsides beside a small fire. Evidently, people gave to His ministry, because we read that He and His disciples had a bag, and that *Judas "...bare what was put therein."* (John 12:6). One time Jesus needed money. Evidently, He and Peter were not with the other a time, or the bag was empty, because Jesus had to send Peter to the sea on a fishing trip to get money to pay their tribute (taxes). The event is recorded in John, chapter 17 and verses 24 through 27. Let's read it together.

"And when they were come to Capernaum, they that received tribute money came to Peter, and said, Doth not your master pay tribute? He saith, Yes. And when he was come into the house, Jesus prevented him, saying, What thinkest thou, Simon? Of whom do the kings of the earth take custom or tribute? of their own children, or of the strangers? Peter saith unto him, Of strangers. Jesus saith unto him, then are the children free. Notwithstanding, lest we should offend them, go thou to the sea, and cast an hook, and take up the fish that first cometh up, and when thou hast opened his mouth, thou shalt find a piece of money: that take, and give unto them for me and thee."

Like all passages of Scripture, this passage contains a number of lessons that are in addition to the lesson we are studying. For example, Jesus is here concerned about offending anybody. But let us pass on to the miracle of creating the coin.

When Peter said, *"Yes",* he was being inconsistent because he had just confessed,

"Thou art the Christ, the Son of the living God." as recorded in 16:16.

If Christ were the Son of the living God, as Peter confessed, and as He certainly was, then He should not be required to pay tribute to the temple, since He was the LORD of the temple. As the Son of the living God, Jesus was a son, and not a stranger, of the King, and so exempt from paying tribute. This is what the Saviour had to remind Peter of when He asked him if the kings of the earth take custom or tribute from their own children or from strangers. Peter answered that the kings of the earth take tribute from strangers, and Jesus reminded him that the children are free, and therefore exempt from paying tribute. This is what the Savior had to remind Peter of when He asked him if the kings of the earth take custom or tribute from their own children or from strangers. Peter answered that the kings of the earth take tribute from strangers, and Jesus reminded him that the children are free, and therefore the Son of God, Son of the King, was not bound to pay tribute.

But the Saviour was concerned about His reputation. Jesus was no free loader. He did not avoid paying His own way, and we ought to be like Him in this respect. Jesus had not paid this tribute since the tribute collector had to ask Peter about it, because it was a debt He did not owe. Nevertheless to avoid offending anyone, He went ahead and paid the tribute.

It is interesting how the Savior elected to pay the tribute money, which was about sixty or seventy cents in our present day currency. He sent Peter to the sea to,

"... cast an hook, and take up the fish that first cometh up; ..."

Now anybody who knows anything about Peter can assume that Peter was more than glad to go fishing, and he hastily turned south toward the Sea of Galilee, which was very close to them - maybe even in sight, and there he caught the fish. In that fish's mouth, Peter found the necessary coin. What is the probability that Peter could have gone to the beach, cast in a hook, and caught a fish with a coin of the right denomination in its mouth? I have caught several thousand fish of all sizes and shapes. I've never caught one with anything in its mouth. I have seen fish that died while trying to swallow a fish that was too big for it to swallow. They still had the half-swallowed fish in their throat and mouth. We all should know that the probability of evolution happening is far beyond the realm of possibility. Evolution scientists have figured the possibility of a single cell coming together that could support life, and found the chances to be completely impossible. The probability of a fisherman catching a fish with a needed coin in its mouth is in the realm of the impossible.

The Creator and LORD of matter was involved in this incident, and He knew what to do and how to do it, and He had the power to do it. Jesus Christ created the coin. He created the fish as well, and He created a fish with an appetite for whatever sort of bait Peter had on that hook. A creative act is a genuine miracle. Many people can create things like drawings, paintings, architectural designs, and even fishing lures, but only the God of creation, the LORD Christ, can create matter from nothing. There are those I meet along the way in Baptist churches who tell me that they do not believe that God created the universe from nothing - they argue that matter was always in existence. They do not realize what they are saying. All pagan religions on Earth that teach anything about creation, teach that their god created the universe from matter that already existed. That is pantheism. That god is in everything, and everything is god.

Tribute money was probably a silvery coin called a "double drachma." The temple tax, therefore, was two drachmae, or a half a shekel. Silver is one of the elements Christ created in the beginning, but now He needed a little more silver, so He created more silver atoms. A silver atom has protons and electrons, and Christ Jesus created these in the right order and motion to form silver. The silver had to be mixed with a small amount of copper as were the coins of the realm, to make the silver hard enough to serve as money. Christ created the copper atoms. The coin had to be formed in the shape of a coin, and had to be inscribed exactly like the coin of the realm.

Some might accuse Jesus of counterfeiting Roman money, but not so. This money was genuine, and made of silver which did not decrease the wealth of the Roman Empire, but rather added a little to its wealth. Caesar would have been happy if Christ had created a truck load of such coins.

Read Exodus 39:26. It is here in Exodus that silver takes it symbolic meaning: redemption. It is here that every Israelite man over twenty years of age had to pay the a ransom charge. Notice that regardless of the age or status of the men, all had to pay the same amount for their redemption.

John 20:30, 31; 21:25

CREATIVE SIGNS

T he LORD Christ performed many creative miracles. When He healed a sick person, there was usually a need for creating tissue that was lost during the illness of the sick one. Often, body fluids are lost during an illness that must be replenished. Physicians do this today through intravenous injections. Jesus would have created those fluids in the body where they was needed. As most other adults, I have had to undergo the misery of having intravenous needles stuck into my hand or arm. For two years now I have had a Groshong CV Catheter sticking out of my chest, left there after I underwent cancer treatment. As life on Earth declines, there are more and more sicknesses of more serious nature. Man's accumulated knowledge allows medical science to continue its advances, and for that I am thankful, but our blessed hope is not in medical science, but in the soon return of our LORD Jesus. We are commanded to pray, *"Thy kingdom come."*

Jesus Christ performed these miracles, not to entertain the crowd, or even to draw crowds, but to demonstrate to a skeptical world that He was indeed the Christ. God did not expect people to believe without good reason. Faith is based on good evidence that the thing believed is true. There can be no scientific proof involved, for there might be a point at which Faith would become sight, and then Faith would be no more Faith. Scientific evidence can never prove or disprove the existence of God for that is in the realm of metaphysics, and science does not function there. Science must confine itself to the realm of matter, and the Laws that govern it.

Science deals with that which can be examined by the senses, which is the soul and body of man. Science cannot deal with the metaphysical, which only the spirit of man can examine. The miracles of Jesus in a sense bring together the spiritual and the physical, for only God can create or eliminate matter. Humans can only believe or disbelieve this. There is no way to prove or disprove by science that God can create. We may use illustrations, but illustrations must always break down at some point; otherwise the illustration would become the fact.

To create a fish, as in the feeding of the five thousand, the Creator must not only create the atoms of matter which compose the tissue of the fish, He must also construct these atoms into molecules of the exact order, He must weld these molecules into the tissue of the fish. In order to do this in the time frame of the event, He must do it instantaneously, as when He created the cosmos. All of the various organs in a living thing are composed of the same atoms of matter, but the molecules of the organs are different. This is not simply superstition garbed in science. The spirit of man can be deceived, but deception can be recognized by the spirit of man if the man becomes so desirous of truth that he is willing to make the sacrifices necessary to find it. In other Words, he must be willing to sacrifice what he has learned in his life if he finds that it doesn't survive the light of reason and objective examination, and the Word of God.

For example, a Mormon could free himself from deception if he would objectively study the Book of Mormon in the light of what is known about American history. A Catholic could free himself from deception if he would have a scientific test done on a piece of bread after a priest claims that he had turned it into human flesh. Deception is usually very easy to recognize if the deceived one will just step back and look at it rationally.

Dr. Isaac Asimov said that he did not believe in God. Yet he candidly admitted that he had no reason to explain his refusal to believe in the existence of God. The most intelligent person on Earth can be deceived. Deception is believing something in spite of good, sound, objective, scientific evidence against it. Deception is irrationality.

In John 20:30 and 31, The Apostle writes, *"And many other signs truly did Jesus in the presence of his disciples, which are not written in this book: But these are written, that ye might believe that Jesus is the Christ, the Son of God; and that believing ye might have life through his name."*

Seven of these signs recorded in the Gospel of John were evidently creative miracles, three of which we have spoken of already: turning water into wine; and feeding the five thousand; and the raising of Lazarus. The other four are John 4:46-54 where Jesus healed the nobleman's son; John 5:1-9 where Jesus healed the man at the pool of Bethesda; John 9:1-7 where Jesus healed the man born blind: and John 21:6 where Jesus created 153 fish in the disciple' nets.

The seventh miracle recorded in the Gospel of John, which is actually the eighth one we are considering, is the instance where Jesus entered the upper room where the doors were shut, as recorded in John 20:19-23. Like the walking on the water miracle, which is not recorded in John, I cannot think of any necessity of matter being created for this miracle. It is important to note that Satan is capable of all sorts of lying wonders, as II Thessalonians 2:9 informs us. Read it now.

The incident in Exodus 7:10ff where the magicians of Egypt cast down their rods, and the rods became serpents, is an example, as well as the passage in Revelation 13:15. Satan's power is only second to God's in the universe.

Satan cannot create life and he cannot create matter. He is a master of deception. He controls the lives of people who desire what he can provide spiritual control over the minds of people. Modern healers are among the most successful deceivers in history. With the technology of television, they use crowd control and mob psychology to create scenes that carry away millions of gullible souls who mistake hysteria for worship.

Popular faith healers can be easily identified as deceivers because they corrupt the Word of God. An example of this is the twisting of the Book of Job. Rather than acknowledging **Job** as a man of great faith as the Bible depicts him, "faith" healers who preach that man was meant to live rich and healthy, make **Job** out to be a man of no faith. The healers say that **Job** lied when he said, *"...the LORD hath taken away,..."* (**Job** 1:21), but God said **Job** was perfect, so you can believe **Job** rather than liars. We must remember that Satan appears as an angel of light. He is seen before great audiences in $800 suits and loads of flowers, or guitars and drums. In the very next verse, the Word tells us *"In all this Job sinned not, nor charged God foolishly."*

- 97 -

In the last chapter of John, Jesus performed one last great creative miracle. It is recorded in 21:6. The disciples had decided to go back to their old occupation after the Savior's resurrection. They worked all night and caught not so much as a minnow. As the sun rose they saw the LORD Jesus standing on the shore. He called out and asked them if they had caught anything, and they answered that they had not. He commanded them to cast their net on the right side of the ship, and when they did, the Creator Christ filled it full of big fish: 153 to be exact - and they were unable to bring them on board. They drug the net to the shore where Peter pulled it up on the bank, and they marveled at the catch.

Jesus then invited them to share the food he had prepared. They knew the Saviour, but not one of them dared to ask Him Who He was. Jesus had created the bread and fish they were eating. The fish was, in my humble opinion, the most delectable food these people had ever eaten, it was plump and entirely without bones. It was cooked to perfection.

The last verse in John's Gospel tells us

"And there are also many other things which Jesus did, the which, if they should be written every one, I suppose that even the world itself could not contain the books that should be written."

The creative miracles set Him apart from the founders of all the world's religions. Only God can create. No religious book on Earth reveals a god who can create. The resurrection of Christ is the sign He gave to the Jews that He was indeed the Christ.

HEALING At MOTHER IN LAW

This is the seventh in our series on the creative miracles of the LORD Jesus Christ, and today we go to the Gospel of Mark for the first miracle recorded there. This is early in the ministry of Jesus, but His fame is spreading rapidly because of His power to cast demons out of their victims. He had just cleaned house with the demons in Capernaum, and the people were amazed by the new doctrine He preached, and the authority He had to command the unclean spirits. All Galilee was a-buzz with the news of the ministry of this Man.

In verses 29 through 31 of Mark 1, the Bible says, *"And forthwith, when they were come out of the synagogue, they entered into the house of Simon and Andrew, with James and John. But Simon's wife's mother lay sick of a fever, and anon they tell him of her. And he came and took her by the hand, and lifted her up; and immediately the fever left her, and she ministered unto them."*

When the mistress of the house is disabled, there is trouble aplenty. Men go hungry and little children cry. How can men and children live without a wife and mother? or a good grandmother? As soon as Jesus entered the house, He was told, probably by Simon's wife, that her mother was sick of a fever. Fevers are usually caused by viruses.

Viruses are pernicious little things that cannot easily be treated by drugs or other medicines, because they live in living cells, and killing the **virus** often means killing the cell. When a **virus** attacks a human or animal, the body raises its temperature in an effort to kill the **virus**. Sick cells secrete interferon which protects the uninfected cells. This protein can mount immunological attacks against particular types of **virus**es. The immune system produces antibodies and sensitized cells that are tailor made to neutralize **virus**es, according to Compton's Encyclopedia.

Viruses cause fevers like measles, mumps, poliomyelitis, and rubella. Treatment of established viral infections is limited to relief for the patient. Few drugs can be used to combat viruses directly. Long term immunization is the best protection against viruses, and short term, or immediate protection, must be administered very quickly after the virus enters the victim, or else it gets a hold that makes it impossible to stop by medicines.

Peter's wife's mother had a fever, but what caused it is unknown, of course. No doubt she was very weak and dehydrated. Man has not known about viruses very long, because they are so small that they could not be found. Dr. Luke probably knew very little about viruses, and if Peter's mother-in-law had a virus, the only thing that could be done for her was to make her as comfortable as possible until she died or recovered.

We know virtually nothing about Peter's mother-in-law. We read this account in Matthew 8:14,15 and Luke 4:38,39, but those accounts add nothing to our knowledge of this dear woman. Since she evidently got up at once after being healed with no complaining or hesitation, and went to the kitchen and began her work, we can assume that she was a good and faithful woman. The workplace of a woman is the kitchen, and her tools are pots and pans, mops and brooms. A man belongs to the cab of a big truck, and his tools: hammers and wrenches, and forklifts and computers, steering wheels and rifles. If America could just remember those good old timey values, we would reduce the divorce rate, and the number of neglected children by thousands; and the juse of dope.

Peter's mother-in-law was concerned that her daughter would not be able to prepare for the guests all by herself, and she lent her skilled hands. Men need women to take care of them, and women of wisdom know that and do what is expected of them. Men are born to wear themselves out and die for the helpless. Women are born to care for the helpless. I am fortunate that I was raised among women who knew what the purpose of a woman is. My grandmothers had no ambitions to be judges or truck drivers, or news idols. They busied themselves in making peach cobblers and fried apple pies for little boys like me who always knew what to do with them. Thank God for noble, godly women who can build men

And please, no mother-in-law jokes. My mother-in-law was a courageous woman who fought a losing battle against the heart break of losing her oldest son in North Korea. He disappeared on a battlefield without a trace in the civilized world, and for seven years she lived in hope, but then the hope was taken away with the notification that her son had been declared dead. Then she slowly died. It has been said that the worst agony of soul is to have a child die before the parent. One can adjust to the death of a husband or wife, because that happens so frequently, and is a part of the human experience. But a child just shouldn't have to die before the parent.

When Jesus came to the bedside, He took the dear woman by the hand, and raised her to a sitting position, and the fever left her at once. Whatever caused the fever was banished from her body. The body fluids she had lost were at once replaced by a creative act of the Creator. The weight she had lost was instantly restored to her. The strength she had lost was restored. She got up, and went to the kitchen and helped her daughter with the work of providing for their guests. Her healing and restoration were instantaneous because Jesus doeth all things well. Every need of her sick body was provided by the miraculous intervention of her caring Savior. God looks upon the world in pity and compassion. God cares for every human upon the earth. If the people upon the earth cared for themselves as much as their Creator cares for them, they would secure themselves in His eternal love and grace. Even the malignant **virus** flees before the power of the great Physcian.

I know not how the LORD Creator Christ healed Peter's mother-in-aw, but I know He did. He knew what was in the innermost cells of her body. He knew every cell, and He knew which ones were sick, and which ones were well. There was no need for x-rays, blood tests, stethoscopes, or blood pressure devices. Jesus knew at once what needed to be done for the sick one.

Jesus knows what ails the sinner. Jesus knows what is in the heart. There is no escaping the truth that Jesus Christ is the Son of God. He is the eternal Creator extolled in John chapter 1 and Hebrews chapter 1. All of the ailments you suffer from, He has already diagnosed, and He can heal them. The wreckage of your life can be building blocks in His hands.

The spectator who witnessed the scene in Pharaoh's court when Aaron threw down his rod, and it became a serpent; and then

Pharaoh's magicians threw down their rods, and they became serpents, would have been deceived. Aaron's rod-serpent swallowed down the rod-serpents of the Egyptians. The enchantments of the Egyptian magicians were so clever that ordinary people could not tell the difference in the miracle and the enchantment. The Egyptians were left with only their senses to make a choice, but Aaron and Moses, knowing the true God, knew the difference.

Not since the Word of God was closed has man needed anything else to support his Faith. Roman 10:17

Most "healing" services are no more than ecclesiastical entertainment for worldly people who have no taste for worship in spirit and in truth; or for desperate people who are so sick they are willing to turn to anything that holds out hope for them. In this we see cruelty carried to a brutal extreme. Giving false hope to the hopeless is nothing less than a work of Satan. Sick people get desperate to be healed. Good health is something that is ignored, but poor health preoccupies our time and thoughts. A well person rarely prays, but the sick person prays frequently. God can give good health or poor, but He is wise in giving either. The greatest honor and opportunity is given to sick folks to be a blessing to God and man.

The body is so complex we are amazed that all the parts of it can work together, even when one part is so sick it malfunctions. Healthy body parts rush to the rescue of the sick part, and adjust their own functions so that they can operate without the assistance of the sick part. Often the body has to function without an organ or part. The body adjusts to the absence of an appendix, or a gall bladder.

CREATED NERVES

In the creative miracles of our LORD Jesus Christ we have now seen Him create wine, **manna**, bread, human tissue, a coin, fish, and body fluids. All of these miracles involved the Saviour in acts of creation that only the true God of the universe could accomplish. Paul understood all of these and much more, and that is why he was able to exult,

"...for I know whom I have believed, and am persuaded that he is able to keep that which I have committed unto him against that day." Paul knew the faithful God. Paul knew the God of integrity Who never violates His own oath or covenant. The Creator stands above and apart from all idols, religions, and gods created by man.

Evidently, there were a great number of people suffering from palsy in the days when Jesus was on the earth. This seems to be a sickness that affected the nervous system so that the person suffering from it could not control the shaking of their arms or legs. It was a debilitating illness that would cause a great deal of fatigue. Probably, there was a loss of nerve tissue, or damage to nerve tissue that made it impossible for the sick one to hold his limbs still. A person suffering like that could not work, and his life would be very miserable.

In Mark 2 beginning with verse 1, we find the account of such a man. Four men brought a sick man to Jesus on his sick bed, but couldn't get into the house where Jesus was, and so they went up on the roof with him, and tore a hole in the roof big enough to let him down into the presence of the Saviour. Jesus honored their the Faith by forgiving man's sins. This man was sick because of sins in his life. This is revealed to us in verse 5 where Jesus said, *"Son, thy sins be forgiven thee."* This caused a controversy with the scribes who challenged Jesus' authority to forgive sins. They asked a good question when they said, *"...who can forgive sins but God only?"* People who sin expose their body to all sorts of illnesses. Great numbers of people are sick as a direct result of their sins.

There is an interesting statement in the 84th Psalm, verse 2.

"My soul longeth, yea, even fainteth for the courts of the LORD: my heart and my flesh crieth out for the living God." That is strange to hear -"...my flesh crieth out for the living God."

When we read about the flesh in the Scripture, we read about that part of the human being which is at enmity with God (refer to many scriptures in Romans), but here we are told by the psalmist that his flesh cries out for the living God! "Flesh" in Psalm 84:2 does not refer to the fallen Adamic nature of man which is at enmity with God, and is the monster in all of us that Paul refers to so often.

What is referred to in Psalm 84:2 is the literal flesh. The tissue which makes up the vehicle we ride around in. The flesh that is kept alive by the blood flowing through the veins. Our literal flesh cries out for the living God because it depends upon the living God for health. Our flesh, including especially our nervous system, needs the blessing of God to remain healthy. When the soul that inhabits a particular body of flesh sins against God, then God's blessings upon the body are cut off. When a person draws polluted smoke into his lungs constantly and frequently, that deprives the flesh of necessary oxygen and that deprives the flesh of its ability to cope with diseases. American Indians had a lot of eye problems because of the smoke of their fires that they often had inside their homes. The pollutants in smoke also cause cancers and other ailments.

When people eat large quantities of food that they know is harmful (gluttony), that is a sin that may result in illness. I believe that many people are sick because they eat food that they haven't blessed by thanking God for it. According to I Timothy 4:4 every creature is good if it is received with thanksgiving. Therefore, we can see that if food is not received with thanksgiving, it may be harmful to us.

Whatever were the sins of the man sick with palsy, we are not told. The reason Christ healed this man was to prove that He is God. Other gods should be willing likewise to prove their deity by healing the sick. Does Allah heal the sick? As I understand it, Allah proves he is a devil by killing innocent people, and making their survivors very sick. Does Joseph Smith heal the sick? Does your god heal the sick? I speak not of carnival shows where there is wholesale "healing." I speak of the gods of religious books. I speak of the gods that people refer to when they assert that all religions lead to Heaven.

Only the Creator can heal because it nearly always requires the creation of new matter. The man sick of palsy had missing nerve tissue, or he had damaged tissue that had to be replaced. Jesus Christ, the Creator, was able to do what was needed to heal this man.

Notice that first of all, the man's sins were forgiven. A sick person cannot expect healing until he has been forgiven for his sins. God will forgive sins when there is a repentant and contrite heart. A sick bed is a good place to examine your life. Don't spend time feeling sorry for yourself, ask God to give you a clear and unmistakable view of your life from His viewpoint. What does God think of you? Are you right with God? Of course, a person who is in a sick bed because of his sins shouldn't have to think too hard about what the problem is.

A sick person is fortunate if he has people around who will support him in prayer. Not only prayer for his healing, but prayer for his walk with the LORD. The man of our story had four friends, and I suppose they have been used for many different illustrations. Their determination is evident in their destruction of their neighbor's roof, and we wonder how they could get by with that. Since Jesus used the plural when speaking of Faith, *"When Jesus saw their Faith..."*, it must be that the four friends knew of the sick man's sins, and were interceding for him in prayer, because it was upon the basis of *"...their Faith..."* that Jesus forgave the man's sins.

Fortunate indeed, is the person who has friends close enough that he can confess his sins to, and gain that friend's prayer support. If one of your close friends came to you and confessed a terrible sin in his life, would you be able to handle it? Or would you be so reviled that you would desert your friend? Herein is one of the failings we have as devout Saints. How many pastors could hear the confession of sins from one of his friends, and then pray with that person for his forgiveness and restoration? Most people are not guilty of being sinfully judgmental, but to know about the secret sins of a friend, and not be judgmental is a very big challenge. We do not have to confess our sins to anybody, except the LORD Jesus Christ to get them forgiven, and confessing our sins to another person could harm that friend, but there are certain circumstances when it is helpful if we have a devout and powerful friend praying for us that we could overcome our besetting sins. If a Saint is having problems overcoming the temptation to drink alcohol, he would be wise to find a kind and devout friend to pray for him, or even be with him, during the times when he is most open to the temptation to drink. The same with any sin. We should bear one another's burdens, and share one another's load of sins and temptations.

The man sick of palsy was forgiven of the sins that had caused his illness, and then he was healed. God will forgive anyone who comes to Him in repentance and Faith, and He will heal and restore when it is His will to do that. Some sins leave such damaged tissue that it would require a creative miracle to heal, such as this man, but God is able to do even that when it is His wise decision to do so. Often, it is not God's will to heal the physical damage done by sin, and in such cases the person must continue to love and trust the Saviour, and depend upon the LORD for sufficient grace. Like **Job** of old, we should have this attitude:

"Though he slay me, yet will I trust in him: but I will maintain mine own ways before him." (Job 13:15).

Nerves can be very sick sometimes. Some people claim to have a "nervous breakdown," but that term is a layman term, and not a medical term. The term was coined by people who cannot explain their illness any other way. Here is what a doctor at the Mayo Clinic had to say about the term: "Daniel K. Hall-Flavin, M.D. says, 'The term "nervous breakdown" is sometimes used to describe a state in which a person is so severely and persistently distraught that he or she is unable to function at a normal level. Nervous breakdown isn't a medical term, however, nor does it indicate a specific mental illness.' " (From the internet: <http://www.mayoclinic. com/ health/Nervousbreakdown/AN00476>

MADE WHOLE

S ome of the Savior's healing miracles were obviously creative miracles, and some are not so obvious, and the Bible does not specifically tell us that any of His miracles are creative miracles, As with many other matters in the Bible, we are given enough knowledge to make certain assumptions which add to our Faith and appreciation of the LORD, and to His everlasting glory, but we are left to wonder.

Our belief that certain of the LORD's healing miracles were creative miracles is the phrase *"made whole"* which appears at least 15 times in the Gospels and once in the Acts. If an individual is made whole by a healing miracle, then it is can be assumed that he was not whole before the miracle when he was sick. This phrase "made whole" is used when Jesus performed all sorts of healing miracles: an issue of blood; diseased; vexed with devils; sick; resurrection; impotent, blind, halt, withered, whatsoever diseases; lame -all were made *"every whit whole."* (John 7:23).

A Gentile woman once came to Jesus distraught because her daughter was *"...grievously vexed with a devil."* What an awful torment this matter must have been to this poor woman. We know not what age her daughter was, but whatever the age, the mother's heart was torn to shreds by the terrible condition of her child. A child is always a child to a parent. We never seem to get over the feeling of parental responsibility for our children. We must allow our children to live their own lives, and make their own plans, and make their own mistakes, but
we can never be completely detached. This woman was a Syrophenician, living in the *"...coasts of Tyre and Sidon."* Since she was a Gentile, Jesus woul not hear her when she called on Him as the son of **David**. He was indeed the Son of

David, but the King of Israel claimed no Gentile as His subject. As the Son of **David**, a Gentile could make no claim on Him, for as Son of **David**, He was strictly the Jewish Messiah. He had already quoted Isaiah's prophecy, 42:1,

"Behold my servant, whom I have chosen; my beloved, in whom my soul is well pleased: I will put my spirit upon him, and he shall shew judgment to the Gentiles."

Messiah *"...came unto His own, and His own received Him not."*

When He sent out His disciples in Matthew 10:5,6, He charged them that they go not unto the Gentiles.

"These twelve Jesus sent forth, and commanded them, saying, Go not into the way of the Gentiles, and into any city of the Samaritans enter ye not: But go rather to the lost sheep of the house of Israel."

Messiah would turn to the Gentiles, but only when the final rejection of the Jews was in. That rejection was the minute the Jews screamed at Pilot, "We have no king but Caesar." God turned to the Gentiles when He called Saul of Tarsus to be the apostle to the Gentiles in Acts 9:15, "...for he is a chosen vessel unto me, to bear my name before the Gentiles, and kings, and the children of Israel."

When, at last, the Syrophenician woman called on Him as "LORD", He responded. His response sounded quite rude. Jesus did not come to Earth to use unnecessary Words. People must know that He is Christ. As Christ it would be expected that He would speak as the situation demands. Nice people compromise. Jesus was nice when the occasion called for it; He was plain when the situation called for it. Here's what He said to this mourning mother:

"It is not meet to take the children's bread, and cast it to dogs."

Most people would have been shocked and offended by that, and they would have left in a huff. This woman believed that this Man was the Christ, the Son of **David**, and therefore, He could do no wrong. Her response:

"Truth, LORD; *yet the dogs eat of the crumbs which fall from the master's table."* is a jewel of humble wisdom and worship.

This woman's great Faith is seen in her acceptance of her position as a Gentile "dog," and her belief that even dogs are

welcome under the LORD's table. We should greatly admire this woman. Her Faith was sorely tested, but she was equal to
the test. She was insulted, but she was willing to understand the insult by the One Who could do for her sick daughter what no one else could do. Jesus, too, recognized her character and her great Faith. At once, her daughter was healed. She *"... was made whole from that very hour."* Creative miracles, and all other miracles, are always instantaneous just as the creative acts at the beginning were instantaneous. Christ spoke and it was done. Talk of billions of years for creation is silly on the college level.

Jesus Christ is the Son of God. He is the Creator and LORD of the universe. Man is wise when he remembers his place in relationship to the Son of God. Christ Jesus is no glorified mortal like so many people in our day strive to make Him. We would do well to approach Him in fear and awe as people who recognize who He is. Christ was the Word Who was with God in the beginning. He was God, and He is God. Numbers of people have more fear of popular evangelists than they do the Son of God.

The little daughter was *"...made whole..."* because she had a momma who had enough wisdom to go to the One Whois altogether able to help. I pity the little children who do not have parents with enough sense to know that it is Jesus Christ Who is the lover of their children's souls. The worst child abuse in America is heaped upon little children whose parents keep them from Jesus Christ. I am convinced that every child would gladly receive Jesus Christ as his Saviour if he were shown the way.

During the good old Sunday school bus days, it was my great joy to be a bus evangelist. I traveled all over working on bus routes, and preaching to boys and girls on Sunday morning. In Maryland one weekend, I preached to about 600 boys and girls, and so many came forward during the invitation we were swamped. Kids learn to love the LORD Jesus when they have the opportunity to know Him. Pastor Casey here in Fritch, Texas where I prepared this essay, showed me a picture of two of his grandsons who are with their missionary parents in New Zealand, and he said that those boys could quote hundreds of
Scripture passages. What fine looking boys they are. At the great Woodland Hills Baptist Church in Tyler, Texas, they have a large turnout of young people on Saturday to hit the streets winning

souls. On Sunday mornings when I am at a church for a conference, I give a program where the children are in front with the adults looking on, these boys and girls thrill me with their intense interest in what I am saying. God still has His "6000 kids."

The tragedy is that so many churches are neglecting to ground their young people in creation as found in Genesis, and when they go off to college they are lost to the rabid evolution professors who pull their brains right out of their heads. A young man in Fort Worth, Texas told me that his sister was a

devout young lady, regular in attending church, but by the time she got her degree from college she had denied her Faith, and would not so much as talk about anything spiritual.

Sometimes I think there must be a special demon that possesses young people, and convinces them that evolution is true and the Bible is wrong. Of course, a demon cannot possess a Saint, but their influence can be overwhelming when they are in a school environment where they possess the professors. It is important for people to be made whole. All of us were grievously wounded in the fall of the human race. Isaiah speaks of us as being wounded from the top of the head to the sole of the foot, and we are covered with *"...wounds, and bruises, and putrifying sores:"* (Isaiah 1:6). What a desperate condition. It is awful when a person has a body that is sick and wounded, but when a person has a soul that is sick and wounded, then he is in a grievous condition.

SIGNS

In the past nine messages, we have been thinking about the creative miracles of the LORD Jesus. The Bible does not say these miracles were creative miracles, in other Words, miracles that required the creation of new matter, but it seems obvious that the conditions that required the miracle required the creation of new matter. If one is satisfied to simply say, "Well, Jesus made the water into wine, and that's all there was to it," then I suppose that is all right for that person if he is being honest. But for those of us who meditate upon God's Word, and desire to know more than what is evident from a passing glance, how Jesus did these miracles is of great importance. The Bible teaches a great deal by implication.

God performs miracles of creation by the Word of man, when it is His purpose to do so. Paul, Peter, Elijah, all raised the dead, which is a miracle in every sense of the Word. These men acted as agents of the Creator when they performed these great miracles. There was a purpose in performing these miracles beside the benefit to the people involved. Miracles are performed as evidence that God is God, and the agent performing the miracle is an agent of God. The agents always made it clear that he was acting on God's behalf.

Virtually all healing requires the replacement of lost tissue. The Word "tissue" means a group of cells that are similar in form or function, and that includes blood and other body fluids. One of the first things a hospital does in most cases is to inject liquids into a patient by intravenous means. Placing the needle in the hand or wrist is a very painful procedure, but is necessary for giving fluids and certain medications. When infection occurs, tissue is lost, and sometimes a great deal of tissue is lost. This tissue is replaced to a smaller or greater extent during healing by natural processes. Blood brings oxygen and the other necessary elements to the wound to form this new tissue. This is a natural process set in motion by the Creator when He made us. It is not miraculous healing because it occurs as a normal function of the body.

A large surface wound is closed with scar tissue which is different from ordinary skin tissue. Scar tissue will dissolve if there is a prolonged lack of vitamin C. What horror stories we hear of soldiers and sailors of old whose wounds open because of a lack of vitamin C. Teeth fall out of deteriorated gums, bones become exposed as wounds open. It is surely awful. Fortunately, such things are rarely seen today. This condition is called "scurvy."

If a person had great healing power, he surely would not confine such activity to the stages of great auditoriums while people were suffering and dying every hour of every day. If a person has the power to heal, then surely, mercy would demand that he burn out his life healing all the sick folk he could come close to. People do not need to heal today to prove the Word of God is true. He would not have to heal to prove that he is a man of God. There would be no useful religious purpose in mass healing today.

We believe that God can heal on the authority of His Word. We believe that a man is a prophet because he preaches truthfully from the infallible Word of God. A man needs no other credentials. The time for giving signs is past. We have a completed body of truth in the Word of God, and the Bible is altogether sufficient for our Faith and practice.

Unbelieving Jews were constantly challenging the Saviour to give them a sign to prove that He was Messiah. One of these instances is recorded in Matthew 12:38-40.

"Then certain of the scribes and of the Pharisees answered, saying, Master, we would see a sign from thee. But he answered and said unto them, An evil and adulterous generation seeketh after a sign; and there shall no sign be given to it, but the sign of the prophet Jonas: For as Jonas was three days and three nights in the whale's belly; so shall the Son of man be three days and three nights in the heart of the earth."

This prophecy of the Saviour was fulfilled when Christ rose from the dead. It may be that some of these very same Jews were saved when Jesus rose from the dead in fulfillment of this prophecy.

We are not specifically told. Jesus called these people an evil and adulterous generation because they were demanding signs from Messiah when He was giving many signs. Their demands were unreasonable. As we have seen in a previous message John said that Jesus did so many signs and wonders that if they were all written in books that the world could not contain the whole. Jesus did signs and wonders in abundance, but they who will not look cannot see. The miracles of Christ were such that they could demonstrate Who He was to a people who were determined that they would not believe. He came to do what no prophet, priest or king in the Old Testament could do. He was what no one else was. Playing religious games with mass healing or pseudo miracles is an affront to the Christ Who alone could perform such miraculous acts. The miracles of Christ Jesus were either creative miracles

or they involved the Laws which govern the behavior of matter. Christ has a unique relationship to His creation, for He alone can fully understand and control it. The ministry of Jesus was no circus put on for the entertainment of a carnal crowd.

Surely the Jews could recognize that when water was made wine, or when a leper was made whole, that there was a creation of whatever made the difference in water and wine, and surely they could see new flesh on the body of the leper. The Jews had to know that Messiah would be God because Isaiah had prophesied that

"For unto us a child is born, unto us a son is given: and the government shall be upon his shoulder: and his name shall be calleWonderful, Counsellor, The mighty God, The everlasting Father, The Prince of Peace."

Such a One could be expected to have all power and might and wisdom. Jesus was the incarnate God of Creation Who would have no limitations upon His person, His being, or His work. He alone inhabits eternity. The absolute wickedness of unbelieving man is revealed in his attitude toward His Creator and Redeemer. To reject God's offer of love and grace, pardon and mercy, is the ultimate sin. To make a mock at God's holiness is a capital crime that only eternity in Hell can begin to settle the account. It is indeed

"...a fearful thing to fall into the hands of the living God."

Divine retribution is just, sure, swift, and certain.

"God is not mocked: for whatsoever a man soweth, that shall he also reap."

At Genesis 1:14 God made the objects in the heaven, the second heaven. These objects either give out light, or they reflect light, otherwise we could not see them. These objects are wondrous things: our sun, the constellations, comets, and planets. It is not said that God created them. He made them. We are told that He spoke matter into being on the first day of creation, but how He made some of that matter into the lights in the sky, we have not a clue. The easiest explanation, and the one that is probably true, is that He simply commanded the matter to gather itself into the lights. We know these objects orbit in the universe in a perfectly orderly manner. All of the stars we can see are less than 6,000 light years away.1 These lights in the sky were created for signs. A sign is an enduring thing. It stys in place for people to see and examine. On the contrary, a miracle an event, seen by the people who are gathered around. The event passes, and dims in people's minds. God alone can make a divine sign, and only God can perform a miracle.

In His healing, Jesus bypassed the normal bodily function of replacing lost tissue which requires much time, and instantaneously replaced lost tissue by divine creation. He did this to establish the truth that He was the incarnation of the Creator of Genesis 1:1, and therefore the Messiah promised by the prophets. After establishing Himself a very God, He then passed this information on into His Word where men could read the record and believe. In John 14 He spoke repeatedly of His Word and His sayings. This eliminated the need for further miracles to prove He is Christ, or that His Word is His Word.

There are many signs all around us that point sinners to Jesus Christ as Messiah and Saviour. Drunkards are made sober by the grace of God; sodomites are made whole by the grace of God; children are happy and secure in homes blessed by the grace of God; all of nature is a sign pointing to Jesus Christ the Creator.

---

1. D. Russell Humphreys, *Starlight and Time* (Green Forest, AR: Master Books, 1996, p.40.

CREATED ANTIGRAVITY

L et's read the thrilling passage found in John, chapter 6 and verses 16 -21:
    "When Jesus therefore perceived that they would come and take him by force, to make him a king, he departed again into a mountain himself alone. And when even was now come, his disciples went down unto the sea, And entered into a ship, and went over the sea toward Capernaum. And it was now dark, and Jesus was not come to them. And the sea arose by reason of a great wind that blew. So when they had rowed about five and twenty or thirty furlongs, they see Jesus walking on the sea, and drawing nigh unto the ship: and they were afraid. But he saith unto them, It is I: be not afraid. Then they willingly received him into the ship: and immediately the ship was at the land whither they went."

Walking on water is not possible because of the force of gravity, and the liquid state of water. In liquid water, the molecules are too far apart to support a weight that is greater that an equal part of water. When water freezes, then the molecules slow down, and come closer together so that weight cannot pass through them. Objects do float on water, you can even float a needle on water, but there are Laws of gravity and displacement that must be obeyed. A man walking on water would certainly violate all the known Laws of physics that would be involved.

When a ship is launched, it displaces, or pushes back an amount of water equal to its weight. Every pound of cargo laid on that ship displaces or pushes back one pound of water. Each pound of weight added to the weight of the ship pushes the ship down a little bit more. If weight is enogh added, the ship will sink so low in the water that water will cover it, and it will sink to the bottom. Any object that is less dense, or lighter than the water it displaces will float while any object that is heavier than the water it displaces will sink. A needle will float for different reasons. A needle floats because of the surface tension of water. Surface tension enables bugs to scoot around on the surface of water. It also explains why water curls upward around the edge of a glass.

When Jesus walked on the water, He must needs create an anti-gravitational Law to control the water while He walked on it. He would have done the same for Peter. Dr. Henry Morris, who has written a large number of **creation** books, says, in listing miracles in John, that miracle number 5 is "Gravity superseded" (John 6:16-21). The Law of energy conservation was set aside as the LORD Jesus created an antigravitational force of unknown nature, enabling Him to walk on the surface of a stormy sea." How grand to know in a personal way the Creator Who can walk on water. This great fear has been the subject of many jokes, but the cold, hard fact is, He did.

The Saviour does not make us privy to the details of His work, because there simply is not room on Earth to contain all the books necessary for that, as John tells us in 21:25. And, if He did, most of us would not be able to understand them. However, we can think about it, and wonder about it, and glorify the LORD in our meditations upon His Word. True science is the study of the matter that Christ created, and the Laws that govern the behavior of that matter. Everything a scientist learns is something that God in Christ did at some time in history. Therefore, everything God did pertaining to physical things, He did scientifically. Without order, there can be no science. Everything God did, He did in an orderly way, *"For God is not the author of confusion, but of peace..."* (I Corinthians 14:33).

The truth that Christ was able to walk on water is evidence that He is the Creator. God created the earth in a state of **darkness** with water covering its surface. Genesis 1:2

"And the earth was without form, and void; and darkness was upon the face of the deep. And the Spirit of God moved upon the face of the waters."

God divided the waters on the second day so that part of the water rested on the outer surface of the **atmosphere** that He made also on the second day. 1656 years later, He precipitated that water to Earth as rain when He destroyed the world with a flood. Water is very unstable stuff, but God is able to control unstable stuff. Consider the Apostle Peter. (and me)

Peter comes across on several occasions as a very unstable individual, but the LORD Jesus Christ used him and made him a very great man. Many of us have to admit that God is a mighty captain to be able to steer us to productive service. We are often wayward, often unstable as water, as Israel said to his elder son, Reuben, as he lay on his death bed. Genesis 49:4,5:

"Reuben, thou art my firstborn, my might, and the beginning of my strength, the excellency of dignity, and the excellency of power: Unstable as water, thou shalt not excel; because thou wentest up to thy father's bed; then defiledst thou it: he went up to my couch."

Reuben would have been in the line to Messiah Christ, but he was unstable. He was like water. Water seeks its own level. Water won't stay put. Water evaporates when the sun gets hot. Reuben could not stay in his own place even with his own father's wife. **Jacob** could not trust the birthright that included the promise of Eve's Seed to a man who was not stable. An unstable man is a loose cannon. Avoid him like the plague.

Seems to me like a drunkard could handcuff himself to his bed with nothing in reach but some food and a telephone when he feels the drinking urge coming on. That would stabilize him for awhile. God's power to deliver from sin is ineffective if a man pulls the plug.

Jesus walked on water so that we could know that He is the LORD of all creation. He made unstable water stable in some miraculous way. He was able even to make unstable Peter walk on the water. Matthew tells us in 14:28 that Peter called to Jesus," LORD, *if it be thou, bid me come unto thee on the water."* Jesus said, *"Come."*, and Peter climbed over the side of the ship and began walking on the sea.

The sea waves were rolling higher than their heads, and even though it was a stormy night, still the figure of Jesus could be seen. Peter had stepped out on Faith, but now he became unstable again, and in fear, began to sink. He cried out to Jesus and Jesus helped him. Even one as unstable as Peter could be helped to walk on the water by the Creator. Living things need stability. The only truly stable thing man can come in contact with is the LORD.

"For I am the LORD, *I change not; therefore ye sons of Jacob are not consumed."*

says Malachi in chapter 3, verse 6. Because God does not change, we can all enjoy relief that we are not consumed, for God has said,

"He that believeth on the Son hath everlasting life: and he that believeth not the Son shall not see life; but the wrath of God abideth on him."

In the same breath in which He secures the believer in His grace for eternity, God condemns the unbeliever to everlasting torment.

When ancient **Job**, who somehow knew the true and living God, extolled the greatness of God, he answered and said that God

"...alone spreadeth out the Heavens, and treadeth upon the waves of the sea.",

and here in John we see that very God in fact walking upon the sea. *"The Words of Agur, the son of Jakeh, even the prophecy..."* of Agur was

"Who hath ascended up into Heaven, or descended? who hath gathered the wind in his fists? who hath bound the waters in a garment? who hath established all the ends of the earth? what is his name, and what is his son's name, if thou canst tell?"

Bless God, I can tell His Name is Jehovah, and His Son's name is Jesus Who is called Christ. Jesus Christ walked right out of the Old Testament into the New. He is Alpha and Omega, the Creator, the One Who treads upon the stormy sea; and binds waters up in a garment. He is God.

We live in a world of things we cannot understand. I find that others are limited in their understanding of the things around us. I see birds outside my window, and I am amazed by the tiny things. How can a hog wallow in the mud, and produce a ham that is scrumptious? How can a thing like a chicken wander around all day, eating anything she finds, and lay an egg I enjoy for breakfast?

----

1. Henry M. Morris, *The Biblical Basis for Modern Science*, (Grand Rapids, Michigan: Baker Book House, 1984), p.83.

CREATED OIL

Having created atoms of matter, God is then able to make from it whatsoever He wishes. Man can take the atoms of matter that God has created and make many things out of it. In Genesis 4:22 we are told that Tubal-cain was making brass and iron, and teaching other people how to make it long before the flood. I have in my possession a replica of an iron hammer that probably was made by a skilled craftsman before the flood.

There is a Law of physics by which Christ maintains the inventory of matter in the universe. The Law is the First Law of **Thermodynamics**, and the Scriptures that tell us that Christ maintains the inventory of matter in the universe are Colossians 1:17 and Hebrews 1:3. Colossians says, "And he is before all things, and by him all things consist." and Hebrews says,

"Who being the brightness of his glory, and the express image of his person, and upholding all things by the Word of his power, when he had by himself purged our sins, sat down on the right hand of the Majesty on high; ...".

Christ is LORD of every atom. Christ, the Creator is still able to create matter when it is His purpose to do so. He has always had that power. He used that power in the case of a starving widow, as recorded in I Kings 17:8-16. The time was during the reign of Ahab, the degraded king of Israel whom Dr. R. G. Lee called "the fat toad that squatted on the throne of Israel." He was the husband of Jezebel, the painted personification of evil.. God used Elijah mightily to perform seven miracles in his life. Elisha, who succeeded him, asked for a double portion of Elijah's spirit, and proceeded to do sixteen miracles during his lifetime.

Because of the wickedness of Ahab, God caused a drought to come upon the land which was to last for seven years. America had better take note that a wicked leader can cause the wrath of God to come upon a land. The drought in Israel was severe, and God sent Elijah to a small stream where ravens fed him with bread and meat. Is there something between animals and God that man knows nothing of? Animals have never disobeyed God, or brought shame and reproach upon His name. The Bible reveals that animals always obey the voice of their Creator, and that they are special objects of His care.

There is a vast difference in animals and humans. This fact is a good reason to reject evolution, for it is the idea that teaches the erroneous notion that life evolved. There is a great difference in the world view of **evolutionist**s and creationists.

I Kings 17:8-17 says, "And the Word of the LO LORD RD came unto him, saying, "... *Arise, get thee to Zarephath, which belongeth to Zidon, and dwell there: behold, I have commanded a widow woman there to sustain thee. So he arose and went to Zarephath, And when he came to the gate of the city, behold, the widow woman was there gathering of sticks: and he called to her, and said, Fetch me, I pray thee, a little water in a vessel, that I may drink. And as she was going to fetch it, he called to her, and said, Bring me, I pray thee, a morsel of bread in thine hand. And she said, As the* LORD *thy God liveth, I have not a cake, but an handful of meal in a barrel, and a little oil in a cruse: and, behold, I am gathering two sticks, that I may go in and dress it for me and my son, that we may eat it, and die. And Elijah said unto her, Fear not; go and do as thou hast said: but make me thereof a little cake first, and bring it unto me, and after make for thee and for thy son. For thus saith the* LORD *God of Israel, The barrel of meal shall not waste, neither shall the cruse of oil fail, until the day that the* LORD *sendeth rain upon the earth. And she went and did according to the saying of Elijah: and she, and he, and her house, did eat many days. And the barrel of meal wasted not, neither did the cruse of oil fail,according to the Word of the L* LORD *ORD, which he spake by Elijah.*"

Many a great and potent sermon has been preached from this wonderful text. A rich widow woman, made poor by the drought which reduced everybody to poverty, had a tad of meal in a barrel and a little bit of oil in a jug, and she was trying to find a couple of sticks with which to make a fire to bake some bread for herself and her son for to celebrate their dying day.

Elijah spoke the Words of God, but the woman looked at him with the hollow eyes of a starved person, and informed him of her desperate condition. God knew all about her, of course, and as in every case, the LORD never calls on anyone to do something without a proper reward. If this woman had not obeyed the LORD, she and her house would have starved to death. Those rich people dining sumptuously every day in their mansions, ignoring God and the bread of Heaven, are starving. Their souls are leaner than that hungry widow in Zarephath. Two sticks – the cross?

I have heard that **moths** will starve to death on pure wool. **Moths** love wool, and will consume a wool garment in a short time, but if there is no dye in the wool, they will starve to death. A human may love cucumbers, as I do, but if he tries to live on cucumbers, he will soon starve to death. God has said that man shall not live by bread alone. There is plenty in America today, yet people are starving because they do not realize that they are more than a stomach. Many people are wrecking their lives because they are spiritually undernourished. The problem in America today is not the White House, but our house. The problem with America's kids is not that they are not allowed to pray in school, the trouble is that they don't hear any prayer at home. Forget about teaching the Bible at school, and start teaching it at home.

Starvation is an awful way to die. I saw a film of starving Africans. One small boy, standing gaunt and naked, became hysterical, and began to scream. Starvation hurts. I saw people in Russia who were seriously deprived of food. Some were war veterans with no arms or legs. It was a shattering experience, and instilled in my heart a compassion for them.

The widow obeyed the voice of the LORD as it came from Elijah, and God responded to her obedience by creating the food they all needed. That meal in that barrel did not run out because God was able to create more as they ate it. The oil in that cruse did not fail because God continued to create oil as they used it. God replenished their supply of food by creating it as they ate it.

Don't blame God for not doing something unless you have asked Him to do something. *"... ye have not because ye ask not.",*

James tells us. Sometimes I think God must get tirder than that appliance man on tv waiting on somebody to call Him. If I have a dollar in the bank, I won't get it unless I ask for it.

The poverty some people live in is pathetic. Great numbers of American homes are completely devoid of peace, joy, and love. Many homes have not a shred of happiness. The children never see momma kiss daddy because momma is in "lust" with some hunk at the place where she works, and she can't stand daddy any more.

The widow's son depended upon her for food. She was his only support. Yet his mother made a cake of bread for a man of God before she made a cake for him. God demands to be placed first in our order of priorities. He must be first. People who do not understand that fact simply do not understand who and what God is. No one should have to explain why God must be placed first above everything and everybody else. There can be nothing before God if life is going to be successful.

God's command to this widow woman through His prophet called upon the woman to put the most important thing in her life behind a man she did not know, and a God she had never seen. Children cannot be first. Wife cannot be first. Husband -whatever - cannot be first. No one else can be served if God is not obeyed. God performed a creative miracle for the widow. I know not what atoms of matter God created to perform this miracle, except the He had to have oxygen, hydrogen, carbon, but people can't live on a night cloud buttered with the east wind (R. G. Lee). There had to be actual substance. But He did not create more than was needed for the next meal. We do not live by Faith to be enriched. Faith will provide our daily needs, but it will not provide for extravagancies. Paul said,

"Not that I speak in respect of want: for I have learned, in whatsoever state I am, therewith to be content."

CREATED BRIMSTONE

B rimstone is fearful stuff even to think about. To know that human souls will be preserved in it for eternity is a horror beyond human expression. Seven times in the Old Testament brimstone makes its awful appearance, even as the breath of God. (Isaiah 30:33) In the Revelation brimstone appears four times to make the heart of man to quake.

Brimstone is **sulfur** which catches fire quickly, and burns furiously with a blue flame. Sulfur is called brimstone because it burns so readily. Rotten **eggs** smell bad because sulfur in the egg unites with hydrogen to make hydrogen sulfide. The sulfur in eggs tarnishes silver very quickly, forming a black substance called silver sulfide. Near the docks in Galveston, Texas there is a virtual mountain of sulfur, because Texas is a leading producer of the stuff, and it is shipped all over the world. Sulfur is a by-product of oil refineries.

It was the LORD God Who invented sulfur. He made it by speaking it into existence just as He created all matter. Sulfur was made for the benefit of man. It is used in a great number of manufacturing processes including the manufacture of tires. When the necessity of a Hell for the devil and his **angels** arose, the LORD made a lake, and filled it with fire which Revelation calls brimstone as we are told in Revelation 19:20 and Matthew 25:41. The gracious LORD never intended for any human to be cast into that awful lake, but many, many people insist on sending themselves there.

In the 19th chapter of Genesis, Abraham's nephew Lot is living in a wretched splendor in the depraved city of Sodom. Lot must have hated it there because the unLawful deeds of the Sodomites *"... vexed his righteous soul from days to day."* God had just visited Abraham, and had told Abraham of His intention to

- 123 -

destroy that pit of iniquity for the unnatural behavior of the inhabitants. God remembered Abraham, and sent Lot out of the overthrow, but nevertheless destroyed the city, and everything in it, including Lot's mocking sons in Law.

In chapter 19 two **angels** visit Lot, sent there by the LORD to get Lot out of the city safely before He rained down the fire and brimstone. These angels appeared as men, and the inhabitants of the city came to the door that night and demanded Lot to send them out for their carnal pleasure. What a shocking thing we hear when Lot actually offers his two virgin daughters to these beasts for their gratification. A sad and revolting story is recorded in Judges 19 of a man who threw his maiden daughter and his guest's concubine to a pack of monsters pounding on his door demanding that his guest be sent out to them for their unnatural behavior. Sodomites are such nice people they get quilts made for them when they die of the natural consequences of their perverted behavior. This is not hate talk, it is the simple truth. Sodomy is accompanied with certain natural deadly results.

At Lot's house the Sodomites demanded the angels of the LORD to be sent out to them, and that was a fatal mistake. The two angels drew Lot back into his house, and struck the brutes at the door with blindness. The next morning early they were incinerated with the rest of the cesspool called Sodom. When unbelievers don't listen, they have to feel.

Frantically, Lot ran to the homes of his daughters, and pled with their husbands to flee the city for it was about to be burned by divine decree. The Bible tells us that Lot *"...seemed as one that mocked unto his sons in Law."* These boys were no doubt successful business men, perhaps financed by the wealth of their rich daddy in Law, and they had that smug self confidence that comes from wealth and success. They had all of Lot's money they needed, and so they gave him the "hee haw." Lot had never spoken to these men about the benefit of serving the LORD, or the consequences of not serving Him. They would have none of his doom's day hysteria now. They were certain that if there were a God Who knew where Sodom was, that He would surely not destroy such a fine city with so many notable citizens. After all, Sodom had the best jobs in the area. There were brilliant men in the city who could produce valuable goods and services. The standard of living was very high, and the average income was far above that of the "straights" who lived in the country round about.

I imagine that the most painful part of all this was that his daughtersjoined in the scorning with their husbands. The grandchildren were well dressed and among the brightest students in the local school system, and even though Lot was a civic leader, he had finally gone over the deep end, and lost his marbles, and was no longer a fit granddaddy to these bright little brats. Morals had no place in the decisions made by these families. They voted for the man who would promise the most pie in the sky, and it didn't matter what he did for personal entertainment. They were ready to accept wealth no matter where it came from or how.

Sadly, Lot walked back through the dark streets to his fine home on the hill. In anguish he paced the floor all night dreading the morning light when his children would be consumed in the fiery wrath of God. But he could not restrain the coming of the dawn. Soon, the gray of dawn was peeping in his bedroom window, and his weary knees forced him to sit on the bed. His wife, who slept like a baby all night because she too, thought it was all a big joke, woke enough to complain to him about shaking the bed, and went back to sleep. Doom hangs darkly over the head of the careless. A little sleep, a little folding of the hands in sleep. So the spiritual sluggard goes to his damnation.

Lot was tired and sleepy, and lay down on the bed to doze, but the **angels** hastened him,

"...Arise, take thy wife, and thy two daughters, which are here: lest thou be consumed in the iniquity of the city."

He lingered, and he procrastinated, dull with sleepiness and hoping that something would happen to deter the wrath of God. Maybe Uncle Abraham could yet persuade the LORD. At last the angels grabbed his hand and the hands of his women folk, and drug them out of the house and out of the city.

"... Escape for thy life;" the angels urged him, "look not behind thee, neither stay thou in all the plain; escape to the mountains, lest thou be consumed."

Lot was afraid to go to the mountains. He begged the **angels** to allow him to go to a small city nearby that would be named Zoar. God's mercy is certainly long suffering. Strange how a person could feel safe in a worldly city without God, rather than on a mountain with God. Should a man move his family to another city just so he can make more money? It's not wrong to want to make more money. A man without ambition is a pretty sorry character. But it's important to remember that the love of money is the root of all evil, and so a man must be very careful about his motives. Elimelech of the book of Ruth moved his family into a heathen land because of a famine, and he did not fare very well. He and both his sons died premature deaths there.

When they arrived at Zoar, Lot's wife looked behind her in spite of the angels' warning, and she turned into a pillar of salt. That in itself was a notable creative miracle, since God had to change the composition of her body from the elements of flesh to salt which is only chlorine and sodium. When the family entered Zoar, God loosed the brimstone, *"...and, lo, the smoke of the country went up as the smoke of a furnace."* Some declare that there is good reason to believe that God burned these cities, but "...that it was in fact, by a volcanic eruption."[1] You are welcome to believe that if you wish, but I prefer to think that God created this fire storm of brimstone by creating the necessary elements and setting them ablaze. God is the Creator, He can use whatever means He chooses. Thank God for Abrahams who have power to call on God for help in time of need.

---

1. Jamison, Fausset, and Brown, *Commentary on the Whole Bible* (Grand Rapids, Michigan: Regency Library, 1961), p. 29.

MADE FLESH

God gave miracles, signs, and wonders as evidence of His eternal being, power, and **Godhead**. It was the miracles that Jesus performed which were evidence that He was the Christ, the Son of God. The importance of miracles in the Word of God cannot be overemphasized.

In a sense every miracle was a creative miracle in some way. God creates things besides matter. He created the Laws of physics, which are commonly called the Laws of nature. In addition to matter, God created energy, force, power, order, information, complexity, biological life, and spiritual life,besides all the things we do not even know about. Unbelieving scientists give several reasons why miracles are impossible. These arguments are very difficult to answer. Most events that are called miracles can be explained by natural reasons. How can anyone prove scientifically that any event was a miracle? or was not!?

Miracles were performed for the primary purpose of giving human beings reason to believe in God or one of His prophets. The benefits of the miracles to the people involved were only secondary. As signs, miracles must be very rare in order to create wonder and get the attention of the persons who hear about them. Folks pay little attention to the commonplace. Overexposure to anything is harmful. Radiation, the sun, cold, heat, etc. are all beneficial when used wisely. I don't think people can be overexposed to God's Word. There is no such thing as overexposure to godly spiritual influences.

"The miraculous can only have significant testimonial value if it is extremely rare. So rare, in fact, as to be beyond reach of the types of rationalization noted previously."[1] (Mass "healers" operate only before great crowds.) "Miracles that can be repeated at the whim of a practitioner, or that can be generated by means of certain specified techniques or incantations are perforce brought within the domain of empirical knowledge by these very facts, and thus are not true miracles at all."[1]

Miracles are performed by the LORD today, but they are not performed today as signs of God's existence or being, or redemptive purpose, and therefore they are not performed for public attention. We know that because we have a completed text of the Scriptures, and man must believe on God and His Son today because of the record God that gave of His Son, namely, the Word of God.

God's primary purpose in dealing with humans is not to maintain their physical health and well being. God will not hesitate to remove these blessings for the spiritual benefit of the individual because "...whom the LORD loveth he chasteneth, and scourgeth every son whom he receiveth." This is the old timey message of God that man so despises today. This is one of the reasons fundamental churches have so much difficulty competing with entertainment-centered denominations. God will bless with good health and happiness those who love His appearance. It is God's will that man be happy and be blessed in this life. It is also God's will that man obey Him and follow His directions. I am not advocating the popular idea of health and happiness for all saved people. I am saying that obedience to God's will is the healthiest way to live. Taking the position that every saved person is entitled to health and happiness as a right is sadly disillusioned. One cannot live as if there were no Master in his life and enjoy life.

The Laws of nature are uniform. This is what makes science possible. Matter and energy are bound by the Laws of physics that God created along with them. The Laws of conservation and decay are basic to the function of the universe. Matter must always be conserved, and entropy must always increase according to these Laws. This is also what makes miracles possible. The only miracle that could be performed in chaos would be to bring order out of it. Miracles were prevalent during certain times in history. During the earthly ministry of our LORD was surely the most important time of miracles. The period of creation was a time of great activity and expenditure of energy by the Creator, but that work may not have involved miracles, if we define a miracle as a reversal or suspension of physical Laws, as I do. God's work of creation may have been altogether scientific. That is why science is the study of what God did in creation. That's the reason a great old scientist said that we only think God's thoughts after Him. Abraham told the rich man in Hell that if his brothers would not believe on the record of the Scriptures, they would not believe even though one should rise from the dead. We can nail that truth down today. People will not believe on the LORD at all, if they do not believe on Him through His written Word. There are no prophecies that there will be another time of miracles before the coming of our LORD Jesus Christ.

For this message, the last one for now on the creative miracles, I have chosen one that I believe to be the most important one in the history of the world. It is an area where we must tread with shoes off, for it is indeed holy ground, And it is intimate ground. We can only meditate upon it from the brief references we are given, but it is a joyful and holy experience to worship God as we ask the Holy Spirit to speak to us and instruct us.

I speak of the conception, gestation, and virgin birth of our LORD Jesus. This was a creative miracle of utmost holiness and mystery. For many years I have thought upon this matter in my deepest moments of worship and adoration of Jesus Christ, but I never spoke of these thoughts because I had never heard or read from. anyone anything that would tend to validate what I believed.

Dr. Henry Morris, in his book The Biblical Basis for Modern Science has written my thoughts so closely that I will just copy excerpts from him. He says. "Actually the miracle was not the birth of Christ, which was a normal birth. in every respect, but rather His miraculous conception in the womb of the Virgin Mary. alleging, that such an event was biologically impossible and thus completely unscientific. Who would have thought the world would use the wonderful works of God in their efforts to annihilate Him?

"There have always been compromising 'Saints' who respond (as they do to other attacks of scientism on the Scriptures) by downgrading the importance of the doctrine and by trying to explain the birth of Christ naturalistically or spiritually." Some have said the divine Incarnation. could have been accomplished merely by an infusion of God's Spirit into the human body of Jesus, without regard to whether His birth had been supernatural or even legitimate. Others say He was the Son of God in the same way all people are

Dr. Morris continues, "All such suggestions, however, are nothing but compromising equivocations, denying the clear record of the Scriptures and dishonoring the unique divine/human nature of the Son of God, destroying the very basis of His great work of salvation. His unique incarnation required an altogether miraculous, supernatural conception, and it is futile, and destructive, even to attempt to explain or justify it naturalistically. It was a biological miracle -in fact. a mighty miracle of creation, fully comparable to the great miracles of creation, when (as it says in Hebrews 11:3) 'the worlds were framed (Greek, karaatizo) by the Word of God.' Hebrews 10:5 says, 'When he cometh into the world, he saith, Sacrifice and offering thou wouldest not, but a body hast thou prepared [same Greek Word, [kaataritzo] me'. "For Christ's body to serve as a sacrificial offering for the sins of mankind, it had to meet two conditions. First, physically it had to be *'without blemish and without spot'* (I Peter 19), carrying no mutant genes (and their physical defects inherited from either parent. 'In Him is no sin' (I John 3:5). Secondly, He must not have a sin nature.

"The only way these conditions could be satisfied would be by special creation of the embryonic body in Mary's womb."2 These matters as so intimate and private that we would not dare address them if the Bible had not given us some details of the matter. We are not able to write as carefully and appropriately as Luke did, but we can nevertheless speak reverently. We must be very careful that we do not allow superstition or old wives' fables surround our thinking and our faith. Nothing could be worse to worship the wrong thing. In Siberia I stood in a Russian Orthodox Church and observed people, especially women, standing before pictures or images with candles in their hands. The poor, ignorant people got carried away into a trance.

1. Henry M. Morris, The Biblical Basis for Modern –Science (Grand Rapids, Michigan: Baker Book House, 1984),p. 80.
2. Ibid. p.386, 387.

THE HOLY SAVIOUR

The Saviour was altogether holy in His life here on Earth as He was as the Creator in Heaven through ages past. Profane humans living in a world of sin, are incapable of understanding fully what holiness is. We must always be careful that our approach to God, and our worship, is always with fear and awe. We must be careful that we do not make fools of ourselves in the presence of the angels. How embarrassing to be found speaking or behaving improperly in the presence of the King of kings. Jesus was altogether a man, and as such we can fellowship with Him, but we must remember that He is also the LORD of glory.

The discussion of the conception and gestation of our LORD is a matter to be approached with the utmost reverence. It is not a subject to be taken lightly, or with mere carnal curiosity, but with a sense of holiness and reverence like Moses must have felt as he stood at the burning bush, and heard God say,

"Draw not nigh hither: put off thy shoes from off thy feet, for the place whereon thou standest is holy ground." (Exodus 3:5)

When the last message closed I was reading a lengthy discussion of the conception and gestation of our Savior from Dr. Henry Morris. It is Dr. Morris' belief, and certainly mine, that the development of the baby Jesus in His mother's body was a creative miracle because it was required that Jesus be holy and without any defects, even in His genes because He was to be perfect. We simply have a passion to know as much about our LORD as possible. I continue to read from Dr. Morris, "Since all genetic inheritance physical, mental, and spiritual -is transmitted equally from both mother and father, it would be impossible for Christ to be born with a blemish-free body and a sin-free nature if either parent (mother as well as father) contributed genes or other genetic materials to His formation. This must be a special creative act of God Himself. 'That holy Thing which shall be born of thee shall be called the Son of God,' the angel told Mary (Luke 1:35). The Saint's Faith rises or falls upon this truth. No religion on Earth makes such a claim as a virgin birth, even though it was foretold millennia before the Savior was born.

"Nevertheless, from the very instant of conception (when His body consisted only of a single cell) on through gestation, birth, life, and death, Jesus experienced a fully normal human life, for He must be Son of man -man as God intended man to be -as well as Son of God. *'Wherefore in all things it behooved him to be made like unto his brethren'* (Hebrews 2:17). That is, He experienced a fully human life in every way, except for sin! Not only did He have no inherited sin nature (Adam and Eve also had no inherent sin), but also, He 'did no sin')I Peter 2:22). He was *'made flesh'* (John 1:14), but it was only in *'the likeness of sinful flesh'* (Romans 8:3). He 'knew no sin' (II Corinthians 5:21). He *'was in all points tempted like as we are, yet without sin'*(Hebrews 4:15).

"Biologically, the Virgin Birth may have been impossible, but after all, that's how we define a miracle of creation, an event that is scientifically impossible but happens anyway!

"Some may have denied this requirement of the special creation of Christ's body, arguing that the absence of a specific genetic tie to Mary would somehow have precluded Him from being truly human or truly Jewish, as the Scriptures required Him to be."[1]

Dr. Morris continues, "But such objections are trivial bespeaking a completely inadequate appreciation of God's ability create! John the Baptist said: *'God is able of these stones to raise up children unto Abraham'* (Luke 3:8). According to the Bible, *'Jesus Christ...was made of the seed of David according to the* flesh' (Romans 1:3), because His legal father was a descendant of **David**, and his biological mother (that is, the one who carried and nurtured him in her womb from the point of conception, and who gave birth to Him) was also a descendant of **David**. In this way, He was made of the seed of **David**. This assertion, however, is not a whit less true because His body was specially formed in Mary's womb rather than carrying Mary's actual genes.

"Nor is Jesus' humanity the least bit lessened by this fact. Adam's body was likewise specially formed (Genesis 2:7) and had no human mother or father. Yet he was fully human; in fact, he was the first man, the prototype man, the father of all men.

"But it is also true that all men who were *'in Adam'* are thereby innately sinners, and it is inescapable that Jesus was *'in Adam'* if he had any genetic inheritance from Mary. Jesus Christ is called the 'last Adam' (I Corinthians 15:45) and, as such, it is not only possible, but appropriate and necessary, that His body (like that of the first Adam) should be directly formed by God. Not only does this not preclude Him from being like Adam, fully human, but it is the only way by which He could be truly human, without sin, as God had intended man to be. At the very least, a special miracle would have to be performed by God on Mary's genetic apparatus, in order to purge the 'sin-factor' (whatever that may be), as well as the accumulated defective mutations of all the generations since Adam. To all intents and purposes, this would amount to a special creation of the newly formed body in Mary's womb."[2]

I was as happy as I could be for years and years, thinking of the Virgin Birth of our LO LORD as being perfectly plausible, ass expected from a Holy God. I was content to think of the night Jesus was born and the **angels** singing, and all the thoughts that go with Christmas and the birth of our LORD. I am still as happy as a child with these wonderful thoughts. But now I am a man, and a man with an inquisitive mind even at 82 long years of age. I wondered how a perfect Man, as Christ was, could be born of a woman who had imperfect genes. To be perfect, the Saviour had to be perfect in His entire Being, and He was that. His Being, of course, including His genes, every cell in His body, and all the rest could not be inspected by man to assure His perfection, but God could. God is the One Who had to be satisfied. There are a great many more elements surrounding the coming of our LORD into the world, especially the truth that He was *"...her seed..."* as Genesis 3:15 promises that He would be. That I still wonder about.

Belief in the Virgin Birth of our LORD Jesus Christ is a matter of conviction, and one of the basics of the fundamental Faith. We understand that Mary was not holy, nor the "mother of God," but that she was a simple Jewish maiden of great faith and devotion to be admired by all. She was a descendant of Adam, Noah, **Abraham**,

Isaac, **Jacob**, Judah, and **David**, and as a descendant of fallible men living four thousand years after the creation, her body would have suffered the ravages of genetic mutations just as all humans have. For her to give birth to a perfect Son Who had none of these mutations in his body, God would have to make every cell in His body as it developed within the body of Mary. Christ was sent *"... in the likeness of sinful flesh ...",* not actually in sinful flesh.

Christ Jesus, the blessed Son of God, Son of man, was altogether and absolutely

"...holy, harmless, undefiled, separate from sinners, and made higher than the Heavens; Who needeth not daily, as those high priests, to offer up sacrifice, first for his own sins, and then for the people's: for this he did once, when he offered up himself." (Hebrews 7:26,27).

Truly did Isaiah write when he penned those weighty Words in 6:1-5

"...I saw also the LORD sitting upon a throne, high and lifted up, and his train filled the temple. Above it stood the seraphims...And one cried unto another, and said, Holy, holy, holy is the LORD *of hosts: the whole earth is full of his glory."*

And Isaiah's reply was perfectly suited to the glorious moment, *"Then said I, Woe is me!"* When a man recognizes his own sinfulness, then he can begin to understand the holiness of God. Dear friends, we are but sinful mortals before such a holy God. Let us come bathed in the blood of God's sacrificial Lamb.

---

1. Henry M. Morris, *The Biblical Basis for Modern* Science (Grand Rapids, Michigan: Baker Book House, 1984), p 388, 389.
2. Ibid. p. 389.

Introduction to Book III

Just as the in the miracles, the book of Psalms have many more references to creation than we can cover in this book. The psalter is the Bible's longest book, and it touches the subject of creation in many passages. I believe creation to be God's greatest work, fourth only to the virgin birth of Jesus, the crucifixion, and resurrection of Jesus. The Psalms address many topics, some topics are given only a Word of two, other subjects get whole chapters. The psalmists praise God from the mountain tops of unrestrained joy, to the valleys of dispair.

Some of the psalms we have chosen for our essays only vaguely refer to creation, and address themselves more to the psalm itself. Remember the Word *"bara,"* for that is the Hebrew Word for God's creative acts. The Word is used in Genesis only three times as God creates matter, the soul, and the spirit. Everything else is made from the matter He created on the first day.

The psalms were written about three thousand years after creation, and they show that people a thousand years before Christ knew the creation story. The creation story had been recorded in the Pentateuch for about twenty-fice centuries. The unpolluted sky at that time must have revealed the universe to the naked eye almost as plainly as the Hubbard telescope reveals it today. People were acquainted with the galaxies and constellations then far more than people are today.

BRIMSTONE

What awful fires rage down in Hell,
That fearful place where nothing's well
Where souls of men are combustible,
And their life sentences are not adjustable.
"If only" echoes through each room
As wretched souls contemplate their doom.
Every soul in Hell believes in Him
Who died upon the cruel cross for them.
Brimstone makes a fearful fire
For foolish men whose sins are dire.
God will not be mocked, He plainly said,
And He will get you when you're dead,
In sweet, cool water dip your finger,
And touch my parch'ed lips – don't linger.
For I am tormented in this flame
Because I rejected God's holy name.

Joseph Kennedy

Book III

THE UNGODLY

In Psalm 1:1 it says, *"Blessed is the man that walketh not in the counsel of the ungodly, nor standeth in the way of sinners, nor sitteth in the seat of the scornful."* Thus begins the longest book in the Bible. Israel's hymn book, it is called, and about half of it written by **David** the great king, who was called the sweet singer of Israel. There are five books of psalms in the Book of Psalms, and all of them exalt our LORD Christ, who is the Creator and suffering Servant Messiah. The psalms curse the enemies of God and His people, and bless believers with joys innumerable. For the next several messages I would like for us to walk through some of the psalms, and from them worship our Creator and Saviour, Jesus of Nazareth, the Lamb of God.

"Blessed is the man that walketh not in the counsel of the ungodly... ." These Words are like *"apples of gold in pictures of silver."* Their worth goes far beyond what anyone can understand until he has lived out most of his life, and either found himself a failure or a success in the great eternal account books in glory.

The Word "ungodly" should be used with care because it is an adjective having God's name in it, and a negative prefix. But God Himself uses the Word, and so it is a Word that must be brought into our vocabulary, but used sparingly and cautiously. There should be no personal satisfaction in calling people *"ungodly"* even though we have the authority to do just that if their life has no place for God. When we call somebody ungodly, it must not be done with malice, though it sometimes can be with anger. When we call somebody ungodly, we must not resort to our personal opinions or prejudices, both of which we all have in abundance. Never call anybody ungodly because it satisfies some psychological hang-up. That would be ungodly on your part.

There are many ungodly people in the world, and a great number of them are nice, kind, educated people who are admired by the world around them. There is probably nobody who would not agree that Hitler was ungodly. Most people would think of Stalin and Susan Smith, the murdering momma, as ungodly people. Pilate, who stood close enough to the incarnate God to breathe His breath, was an ungodly man.

Most people would consider drug pushers ungodly. How about the gambling addicts who throws away the family fortune at gambling tables? Most people would agree that people like these are not godly people. Religious leaders without Christ are ungodly. In fact, it's hard to imagine someone more evil than one who will deliberately lead a person to perdition.

But if ungodly means "not godly," then the number of people we think of as ungodly will have to vastly increase. God is unto a man whatever the man makes Him. When a man makes his boat the most important object in his life, the boat becomes his god. The ungodly make Him nothing. By their life and labor, they seek to convince others that God is nothing to be concerned about. This is their counsel. His counsel says that man should live as though there were no God. The number of the ungodly is exceeded only by the number of people listening to the counsel of the ungodly.

God was so wise in His terminology here as everywhere in the Scriptures. Note that He does not say, "Walk in the counsel of the godly, because one could walk in the counsel of the godly, and also walk some times in the counsel of the ungodly. There would be a great number of people who would never find a godly man. The world's problems can be attributed to the ungodly, and those who walk in their counsel. II Corinthians 6:7, for example, *"Be ye not nounequally yoked together with unbelievers: for what fellowship hath righteousness with unrighteousness? and what communion hath light with darkness? And what concord hath Christ with Belial? or what part hath he that believeth with an infidel? And what agreement hath the temple of God with idols? for ye are the temple of the living God; as God hath said, I will dwell in them, and walk in them; and I will be their God and they shall be my people. Wherefore come out from among them, and be ye separate, saith the LORD, and touch not the unclean thing; and I will receive you."*

It is better to walk alone than to walk with the ungodly.

When God said *"Walk not ... "* the command was clear. Never be so foolish as to walk in the counsel of the ungodly. It is a tragic truth that when a child goes to a state school, he is almost certain to be required to walk in the counsel of the ungodly. The state can never ask an applicant for a teaching position if they are saved. The state doesn't have the authority or the political right to ask that question. There are so many ungodly people working in the education system in America that it is not far from being exact to say that the system is ungodly. Certainly when those who are in control of the educational system in America are considered, it becomes more likely to be true that the American school system is ungodly. I was a Saint working in it, and I felt it was ungodly. Part of my pay was from beer taxes.

The children who must walk in the counsel of the ungodly at school are more than likely to go home to parents who never open the Bible and teach them the blessed truth about creation and salvation. Not only is this true for ungodly homes, but it is true for homes of many saved people. The ungodly have the children at school, and the ungodly has them at home. The school years pass so swiftly that most parents don't even realize their children are in school before graduation day. Children are not helped much by good intentions. Throw in the television and computer, and the situation for a child is almost impossible. How can God's people touch the lives of such people?

Godly parents who must send their children to state schools must take extra care in teaching their children how to learn from the ungodly without walking in the way of the ungodly. Great wisdom is required for a parent to walk on this water. But it is not so much what the parent says to the child that is the antidote for the poison of the state schools; it is how the parent behaves. A parent who is obviously very much devoted to the LORD Christ will have more influence on the child than the parent who counsels without parallel behavior.

H. S. Lipson, professor of physics at the University of Manchester, U.K. wrote, "In fact, evolution became, in a sense, a scientific religion; almost all scientists have accepted it and many are prepared to 'bend' their observation to fit in with it." As a religion, evolution is opposed to the fundamental sanctity that demands strict adherence to the creation record in Genesis which lays the groundwork for the Gospel of Jesus Christ. Without Genesis 1 -3, the rest of the Bible does not make sense. **Evolutionists** are devilishly sly enough to know that while many Saints have never so much as thought about it. Certainly, Saints who support evolution by neglect and lack of faith, have not considered the fact that they have somewhere along the line walked with the ungodly.

Psalm 64:2 offers good counsel in this respect:

"Hide me from the secret counsel of the wicked; from the insurrection of the workers of iniquity."

Our children must be taught that an appeal to the LORD God is always the best action to take when exposed to the dangers of the counsel of the ungodly. Children must of necessity be under the influence of the wicked if they must attend state schools, if not from the teachers, certainly from classmates, and the curriculum, but they need not walk in the counsel of the ungodly. They must do whatever honest thing they must do to get through, and finish their education, but they need not walk in such counsel. Just because teachers heap evolution, and anti-Christ dogma upon them, does not mean that they must heed such counsel or walk in it.

The duty of the Saint, whatever his age, is to pray for those who despitefully use us, and that is definitely what the ungodly do when they counsel us to evil. The student who is praying that God will deal with his ungodly teacher is less likely to walk in that teacher's evil counsel. **Job** was a righteous man, with wisdom that we would do well to heed. He said, "Lo, their good is not in their hand: the counsel of the wicked is far from me." (21:16) The reason the counsel of the wicked was far from **Job** was because he had resisted the temptation, and it fled from him. The ungodly are insulted and offended when their evil counsel is resisted, rejected. They usually do not persist too long. They will retreat and another will take his place. The counsel of the ungodly is an attack that will never be fully defeated.

HEAVENLY GLORY

The study of the psalms raises our concept of God, and motivatesour dull heart to worship Him as LORD, Creator, Saviour. The value of our praise is whether or not it is as real when we are alone as when we are in a great crowd of people. If our heart can worship the living God with such glory that it nearly bursts when we are alone, then we can be sure that we are not carried away with crowd psychology, or carnal emotion, and our worship is worthy of the God and Father of our LORD Jesus Christ.

Psalm 19 is another of the many psalms that bring our attention to God the Creator. I memorized this psalm during my truck driving years, and found it exceeding joy to recite it to myself while the engine of the truck contributed anthems my heart converted to praise. Sometimes. That was usually during the long nights when two o'clock in the morning was little different from two o'clock in the afternoon, when the brightness of His glory shone in the cab of the truck like the Shekinah glory. There is a song: "My God and I go in the fields together … ." I'd sing, "My God and I go on the road together." Sometimes I would open the mike on my CB and sing, and what a commotion that would cause! The truck drivers couldn't stand it.

One day I was in Plymouth, Michigan, and heard two truck drivers cursing each other on the CB. The racial slurs were awful from both of them. I opened my mike, and began to sing, "Jesus loves the little children, all the children of the world." When I finished the little chorus, a woman came on the CB and pleaded with me to sing it again. I couldn't do it, though, because I was then in traffic such that I had to put the mike down to change gears; and turn the wheel. Psalm 19:1 says, *"The Heavens declare the glory of God ... ."* I have discussed the first part of that verse, and showed how that the three-dimensional universe was created by God in such a way as to illustrate His triune Being. A line is an one-dimensional figure. An actual line is invisible. It is only a concept in our minds, and what we draw for lines are only representations of that concept. A plane is a two-dimensional figure, also invisible for the same reason that a line is invisible. A cube is a three-dimensional figure which is visible. The Holy Spirit is like lines which are invisible and fill the universe. The Father is like a plane which is also invisible. But the cube is like Christ Who is visible. The visible cube is made up of twelve line segments and six planes, which shows how that Christ is the manifestation of both the Spirit and the Father, and how that God is a triune God

    "The firmament sheweth his handiwork." The **firmament** is everything from the surface of the earth to the outer reaches of the **atmosphere**. The stratosphere is above the troposphere in which we live. It lies about eleven miles up, and reaches up about thirty-one miles. Of course, nothing lives up in the stratosphere, but nevertheless, it shows God's handiwork. It escapes me how it could have shown God's handiwork in the days when the Words of the Bible were written. Those folks probably didn't know it was up there. But we know plenty about it, because we have learned that the ozone layer is in the stratosphere, and Ozone is what protects us from the dangerous ultraviolet radiation that comes from the sun. Everything therefore that is exposed to air is the result of God's handiwork. I like that Word, "handiwork." It seems to me to speak of great skill. The work of a skilled, careful craftsman is his handiwork. There is another layer of atmosphere above the stratosphere, and there, in the ionosphere, is where the aurora takes place. The ancients could have been aware of this phenomenon glorifying God's handiwork. The firmament is a very complex area. We have learned much about it through modern science, and the more we learn, the more reason we have to glorify God's handiwork. It seems the more reason man has to glorify God, the less glory man gives to God. Just as the more sophisticated our electronic sounds become, the worse man's uses of it becomes. It's a terrible world – made so by sin, and man's impudence.

Evolutionists have written many hundreds of books on evolution. They have conjured every possible notion for a footing, but they ignore the contents of the first page that should be placed in every one of their books. That page says, "There are no transitional forms in the fossil record." As page number one of the evolutionists' books, the rest of their books are worthless. Another thing page one says is, "Not one soul on Earth has ever discovered the age of a rock." Page one must say, "Nobody has ever seen one creature turn to a different kind of creature, nor has anyone ever seen anything that seemed to be one creature turning into a different sort of creature." After reading page one of the evolutionists' books, then one can skip the rest of the books, and turn his attention to the Book that begins *"In the beginning God created the Heaven and the earth."*

Psalm 19 begins with a statement that makes the rest of the psalm beautiful and authentic. How difficult it is to describe Heavenly things in earthly Words. The Heavenly must be carried to the ear by the earthly, but it is carried by the potent skill of the Holy Spirit. As Spurgeon desired, let us also desire to drink at the celestial well, and learn to utter the glory of God!

"Day unto day ... ." verse 2 begins. Day after day the Heavens declare the glory of God. Days could not speak to us any plainer. Even in an office cubicle with no contact with the outside world other than a telephone or a computer, the day speaks to the human organism. This day brings you closer to a face to face meeting with your Creator. What are you going to do about it?

Verse 3 shouts *"There is no speech nor language where their voice is not heard."* The day speaks in every language and dialect on the globe. Headhunters in New Guinea hear the voice of the day as it uttereth speech declaring that God is God and beside Him there is no other.

Verse 4 says, *"Their line is gone out through all the world."* God's telephone lines are never knocked dead by lightning or wind. When I was in the Air Force I saw C-47 transport planes that were rigged to lay telephone wire. These things could fly over one command post, and begin unreeling wire, and in a few minutes, have that command post connected with another. With the line goes Words, and Words are the power of humanity. The power of humanity to build it or destroy it. With the line of God goes Words, and the Words of God are His power.

- 143 -

Verse 7 tells us that *"The Law of the* LORD *is perfect, converting the soul; the testimony of the* LORD *is sure, making wise the simple."*

The Law of the LORD to us is the whole counsel of God. It's perfect. Our poor preaching adds nothing to it, but without it, our poor preaching is merely the raving of idiots. We should be careful in our opinions of other people. We must not accuse anyone without thoughtful reason. No one is anything without the Law of the LORD. It is the Law of the LORD, the Gospel of Jesus Christ, the record that God gave of His Son that is the power. God forbid that anyone should ever forget that, and God help sinners to understand that.

"The testimony of the LORD *is sure, making wise the simple."* It is not IQ that will get a sinner through the pearly gates, nor the good heart of a rich man, but the testimony of the LORD. It is sure. It is certain. It makes the simple wiser than the scholar. Put God on trial. Call all the witnesses you like. Call every **evolutionist**, from the most careful to the most rabid. Question them all. Appeal to all the evidence you can rake and scrape that there is no God. Holler and whoop. All the defense will call is the testimony of the LOR LORD D. Let the day speak. Give the stars a voice. Call Satan himself demand of him, "Did you stand before God? Did you see God create the universe?" He will lie, but he was there, and he saw it all.

It's painful to skip verses, but Time is pointing at the clock. Go to verse 14.

"Let the Words of my mouth, and the meditations of my heart, be acceptable in thy sight, O LORD, *my strength, and my redeemer."*

Of what value is life if a man walks not with God? What value is a brain if it is not filled with the Words of God? How can a heart love if it has no place for the LORD of Love. Keep your coins of lead; bring to God currency molded in the furnace of a heart fired with worship of the living God. The psalm begins with the heavens; it ends within our heart where the Redeemer must reside. Be wise and prudent. Listen to the testimony of the days.

THE VOICE OF THE LORD

The psalm of *"... the voice of the* LORD.*"* is Psalm 29. That phrase occurs seven times in these eleven wonderful verses. We will hear more of the voice of the *the* LORD in a few minutes. The psalm begins, *"Give unto the the* LORD, *ye mighty, give unto the the* LORD *glory and strength. Give unto the the* LORD *the glory due unto His name; worship the the* LORD *in the beauty of holiness."* Now where, I wonder, am I going to get glory and strength to give to the *the* LORD. Certainly, I have no glory. I am a simple mortal with limited strength, with very little in liquid assets, with hardly a talent. I cannot so much as give glory to one of my grandchildren. Yet God commands me to give glory and strength unto Him. God has never commanded me, or anybody else, to do something we could not do. He is too gracious for that.

How do I obey, dear *the* LORD? A child could know. I give God His own glory. All glory is His. I attribute all of His glory to Him. I return all of His glory to Him. When I see the glittering stars in the star-spangled Heavens, I do not attribute that glory to dead matter, or natural selection, or the big bang. I give that glory to God from whence it came. When I see the delicate blossom on my Cana lilies, I realize that thing is the artwork of God, and I return to Him the glory for such things of beauty because from Him it sprang. Everything I behold, and everything I hear and see that is beautiful and miraculous, I return to God the glory for it. The glory does not belong to education or wealth or to the richest of men or the most righteous, or the fairest of women. The glory came from God, and to Him I return it. I give Him His own glory.

When I reflect back to God His glory, I am bathed in the light of that glory. When I reflect God's glory to Him, I am blessed

- 145 -

and filled with rapture by that glory. I worship the **the** LORD in *"... the beauty of holiness."* People proclaimthat the human body is beautiful. Excuse me, I know what you mean, but I think the human body is beautiful when it is clothed in holiness. Holiness is worn over a modest covering. Children come to church today dressed in the same old clothes they wore out to play. Would they go see the president of the United States dressed that way, if the president of the United States had the character of, say, Harry Truman or Ronald Reagan?

Would the daughter of the deacon who came to church last night dressed in short shorts, dress up to go to a wedding or a funeral? Why no. She might dress up to go to a country club party, though. Would she dress up to stand in the presence of God? All I can say is that she didn't last night. No one offers glory to God unless he is modestly and respectably dressed. Check out Exodus 20:26.

Giving God His glory gives a person a sweet spirit of humility. Yes, a humble man can run a big company or the nation. Moses, one of the greatest leaders who ever lived, was the meekest man alive at the time, and "meek" is a stronger Word of self-abnegation than "humble." "Glory to God" will be emblazoned on the coat of arms of a victorious king.

Verse 3 reveals the God of the storm. Years ago when I was in college, I sold cars for a while. One of the leading salesmen was a disabled fellow who was an outstanding salesman, but one of the most profane people I ever met. He cursed God with nearly every breath. There was no fear of God before his eyes until the storm came. When the tempest blew, and the wind bore bullets of rain, and the sky was rent with lightning, he went into a fit of fear. He rolled up no car windows because he was too terrified to get outside.

The chariot of God rides upon the black wind of the storm. A raging sea is a testimony to the power of God, and his rage against sin. The lightning foretells the fearful judgment of God against sin and sinners. The winds demonstrate the mighty power of God.

When I was a lad, a storm came one Sunday afternoon, and Dad was home. We sat quietly while the storm raged, but then when it was past, we got into the car, and drove around the city to see the tree limbs blown to Earth, and now and then a roof was gone. We lived in east Tennessee where tornadoes were not often seen, but summer storms were fierce. In spite of the clutter of the storm, the air was left fresh and cool. It was so refreshing to go out after the storm and look at the clear blue sky, and hear the birds rejoicing that the storm was gone.

We are literally in the hands of God when the storm rages. My Dad always told us very calmly that the storm was the *the* LORD 's business, and we need not be afraid. It is depressing to hear what parents tell their children about eternal things like death. In a misguided effort to make the shock of death easier for little children, some people tell the little ones things that are the stuff superstition is made of. When disaster strikes, counselors are called to the place to counsel those traumatized by the event, but the counselors often know less about the eternal than the kids they are supposed to counsel.

Verse 5, *"The voice of the the LORD breaketh cedars."* I walked with my family along the top of a ridge one Sunday afternoon when we came across a massive chunk of an oak tree trunk. White oak is very substantial stuff, but here lay half a tree along the path, and we looked here and yonder to see if we could find from whence the thing had come. We walked a hundred feet along the trail when the boys shouted that they had found the tree. There stood the wounded remains of the tree slowly dying. One has written, "Black from the stroke above, the smoldering pine stands a sad shattered trunk." The cedar of Lebanon is the grandest of trees, but it is nothing when the lightning reaches down and touches its crown. The most stalwart of men fall beneath the stroke of the Gospel of Jesus Christ. Men who were saturated with sin, and boasted of their hatred of God and His people, have been reduced to weeping children by the lightning of the Gospel. 0, God, that we could see such power again.

Spurgeon wrote, "The voice of our dying *the* LORD rent the rocks and opened the graves: his living voice still works the like wonders. Glory be to his name, the hills of our sins leap into his grave, and are buried in the red sea of his blood, when the voice of his intercession is heard."

Verse 7: *"The voice of the LORI) divideth the flames of fire."* There is a filling of the Holy Spirit for everyone. Baptized, or filled with the Holy Spirit when we are saved, we need constant refilling as we pour out of our springs of life the Word of Life, to others. At Pentecost, cloven tongues like as of fire came as a mighty rushing wind, and sat upon each of them.

Verse 8: *"The voice of the the LORD shaketh the wilderness; the the LORD shaketh the wilderness af Kadesh."*

Desolate is the wilderness of the sinner's life. A life lived for the pleasures of the day. A life of days fleeting by so rapidly that each one should be approached with a sense of awe and reverence. Only the voice of God can shake those in the wilderness of sin, and bring them to the ligh0t of day.

Dr. Morris believes that this whole psalm is more about the flood of Noah, than of an ordinary rain storm. He points out on page 40 of his book, *Sampling the Psalms,* that the Word "flood" in verse 10 is the same Hebrew Word as the one that God used for flood in Genesis 6 - 9: and that this is the only place in the Bible where this particular Word for flood is used otherwise. All of the phenomenon spoken of in the chapter could be a description of the flood. Verse 3 speaks of *"many waters"* and the horror of the inhabitants of Earth is impressed upon us as we again contemplate the first rain from the skies.

When God broke up the **canopy** that enclosed the earth, and precipitated that water to Earth as rain, the **atmosphere** must have greatly expanded and the air pressure on People must have been gasping for breath, and some perhaps dying because they could not breathe. The *"men of renown"* who had exulted in their strength, suddenly found themselves unable to run more than a few steps. The amount of water in the air would have added immensely to people's difficulty in breathing. Some may have drowned standing in the rain.

"And I looked, and behold a pale horse: and his name that sat on him was Death, and Hell followed with him. And power was given unto them over the fourth part of the earth, to kill with sWord, and with hunger, and with death, and with the beasts of the earth."

It seems that Satan is conditioning mankind now through tv to endure in a perilous time

ABSOLUTE JUSTICVE

Another of the creation psalms is Psalm 9, speaking in verse 1 of His *"... marvelous works."* I have been able to bring to my computer screen some interesting things since I went on-line. The most striking revelation about the human race that the internet affords, is, of course, the total depravity of man, and his total lack of wisdom. These are people who are sophisticated enough to use computers, not the ordinary run of nitwits. This raises the question of who runs computers. There are nit wits natural born, and there are nit wits by choice. The natural nit wit deserves our understanding. The self-made nit wit runs computers.

The other night I discovered a very interesting thing. The title of the publication is "What is Evolution," and is found on the internet in "The Talk Origins Archive." The piece is written by Larry Moran who claims to be a scientist, but that is all I know about him, except that he is a strong proponent of evolution. He is evidently a biologist. In his article he attempts to define evolution, and make it clear what evolution means, but he seems to have his own definition, which is not what most folks think of as evolution. Most people think of evolution as being what the Oxford Concise Science Dictionary says it is. This dictionary defines evolution like this: "The gradual process by which the present diversity of plant and animal life arose from the earliest and most primitive organisms, which is believed to have been continuing for the past 3000 million years." That is what most people think of when they think of evolution.

However, Dr. Moran takes issue with that definition, and says it, "... is inexcusable for a dictionary of science." Dr. Moran must be a mighty prestigious scientist to take on the Oxford Concise Science Dictionary. Is it possible that evolutionists are attempting to change the definition? It can be said with a good deal of certainty that under the definition in the Oxford dictionary, there is no such thing as evolution theory.

Moran continues his criticism of the Oxford dictionary by writing, "Using this definition it is possible to debate whether evolution is still occurring but the definition provides no easy way of distinguishing evolution from other processes. For example, is the increase in height among Caucasians over the past several hundred years an example of evolution? Are the color changes in the peppered **moth** population examples of evolution? This is not a scientific definition." Moran is looking for a hole for to hide evolution in. Evolution, being thoroughly discredited scientifically, he now attempts to change the definition.

If the new definition of evolution makes what happened to the peppered moth evolution, then there is no quarrel between **evolutionist**s and creationists. The peppered **moth** lives in England. There are light colored **moths** and dark colored moths, the dark **moths** being less than 2% of the total number before about 1848. Then came the industrial revolution, and the smoke of coal fires filled the air with soot. The percentage of dark colored **moths** greatly increased until by 1898, about 95% of the moths were dark colored. The moth population turned dark because birds were eating most of the white ones because the white ones could be seen better on the soot-darkened trees.

Moran states that this is an example of evolution. If that is an example of evolution then we welcome the evolutionists to the Creationist camp where sanity reigns. Creationists have understood adaptation for a long time. Actually, these moths did not even adapt. There were two colors of moths all along, and the environment made the difference in how many of each there were at any given time, not evolution. The moths were exactly the same all the time.

I am certain that Dr. Moran would be very hostile toward my primitive notion of evolution, and it is primitive, since I am a teacher, and not a scientist, but I have enough sense to understand a few things.

One of the things I understand is that the sciences, biology and paleontology especially, would be greatly enhanced if there were not so much prejudice - vicious prejudice - I might add, against the inclusion of students in universities who believe in a young rather than an old Earth. The Earth is about six to ten thousand years old, but compared with three or four billion years, a six thousand year old earth is just a diapered infant. By their religious prejudices, **evolutionist**s exclude some of the best minds in the world, namely the born-again minds of young Saints who believe the Bible.

How dull science must be when it is approached by people who see nothing but dead matter. When the Creator is given His rightful place as the Source of life, then science vibrates with the joy of life. Matter really is not dead. It was created by a living Creator. What a wonderful thought to take into a laboratory! I doubt that the greatest of scientific achievements, those that founded today's disciplines, could have ever been accomplished if those great scientists of old did not consider themselves as just "thinking God's thoughts after Him."

Today's scientists consider themselves far more sophisticated than those old scientists of the 16th, 17th, and 18th centuries. The modern scientist thinks he has made a great leap forward in placing God in some category other than where the Bible places Him, and where those wise old scientists placed Him. There is no God other than the One revealed in the Bible, and what He revealed about Himself in His Word can be enhanced by the study of science. After all,

"The Heavens declare the glory of God; and the firmament sheweth his handiwork."

Surely, a scientist who is also a Saint should know the *the* LORD better than anybody, except perhaps the preacher.

God's works are marvelous, **David** declared, and, as a result of his understanding that, he could praise the Creator with his whole heart. Psalm 9:1 says, *"I will praise thee, O the LORD, with my whole heart; I will shew forth all thy marvelous works."* The examples of these marvelous works are without limit, from the living eye, to the peppered **moth**, to the lowly fungi. God's works are marvelous in their creation. Man's conception of God is so devoid of Scriptural enhancement that he can't believe in a God Who can create by speaking. Man's natural conception of God is that God must have millions of years to make a living thing. This notion is as primitive as the idea that there has to be a different god for every task: a god to create the stars, a god to create the earth, and so on.

In verse 20 of our chapter, it says, *"Put them in fear, O the LORD: that the nations may know themselves to be but men."*

There are all sorts of **evolutionist**s, with diverse notions about how it all started and where it will all end. The Word from God is that man must remember that he is but a man. What can he know, really, when the total of knowledge is considered? The best thing man can do is to prepare to meet his Creator. This is more important than creating life in a test tube or finding life on some other planet or finding the missing link. A man may accomplish all those great tasks, but if he dies without doing all he can to secure his soul, then he will have lived in vain, and it would have been better for him if he had never been born. The most marvelous work of Jehovah was in bringing to fruition all the prophecies of the Old Testament in the death of His only begotten Son on the cross. In His perfect work, Jesus Christ took from mankind the grievous burden of works for salvation. Man has not the Holy Spirit's power until after he is saved, and the only way man can do any acceptable work in God's sight is through the power of the Holy Spirit. Man alone is incapable of doing good works.

Marvelous works are recorded in the Sacred Text. A needle can be made to float on water. I have done it. Nothing to it. But think of a man making iron float (II Kings 6)! The sun stands still! Fleece is made wet while the ground all around is dry (Judges 6:37)! A man builds a barge 450 feet long (Genesis 6:15)! float when he is standing on the bank of a river (II Kings 6)! Think about an ass talking (Numbers 22:28)! Supernatural Book, this Bible! Omnipotent God! Omnipresent God! Omniscient God! *the* LORD, Jehovah, Elohim, Jesus, Christ, Messiah!

Man is a heroic creature, nevertheless. There are thousands of acts of heroism recorded in human history. Men and women, boys and girls have given their lives to rescue other people. Such heroic acts must touch the heart of God, yet He has spoken. He has said that there is no salvation without the shedding of the blood of an innocent sacrifice, and He was referring to His only begotten Son, Whose blood alone is potent enough to redeem a soul.

GOD'S POSSESSIONS

In Psalm 24:1 we read, *"The earth is the LORD's, and the fulness thereof; the world, and they that dwell therein."*

Evolution is an evil philosophy that contradicts the Bible and its revelation of how the universe and the life in it began. Science is not our enemy. Those who hold to the Word of God as infallible have no quarrel with the scientific work of scientists. Biologists and all the rest contribute mightily to the comfort of humanity. We commend them for this. Our contention is with those who go outside the realm of science, and make statements about the age of the earth while claiming to speak as a scientist. The evolutionists mutter about creatures turning into different creatures over millions of years as they withhold whatever evidence they have for making such preposterous claims. Evolutionist are the modern alchemists.

It may be a fact that a scientist holds in his hand a fossilized bone. No one could controvert that. It is not a fact that that fossil is evidence of evolution. It is a fact that there is a fossil bird called "Archaeopteryx " but it is not a fact that this thing has anything whatever to do with reptiles turning into birds. It is a fact that dark colored peppered **moth**s became more plentiful in the 19th century. It is not a fact that that phenomenon had anything to do with evolution.

I believe I have as much understanding of science as the average person. I understand that water is composed of two atoms bound together in a molecule containing two atoms of hydrogen and one atom of oxygen. I also have a tad of understanding of the human mind. For example, I can detect the fallacious nature of a statement like, "When someone claims that they don't believe in evolution they cannot be referring to an acceptable scientific definition of evolution because that would be denying something which is easy to demonstrate. It would be like saying that they don't believe in gravity!" ("What is Evolution?" p. 3) If evolution is as easily demonstrated as the Law of gravity, then let's get on with the demonstration, and stop making irrational assertions.

Creationists are accused of not understanding what evolution means. Some claim that the dictionary definitions are wrong. If dictionaries are wrong, then how will we know what we are talking about? Certainly, it seems, the **evolutionist**s do not believe we know what they are talking about.

Here is a definition Dr. Moran, quoted in the last essay, declares to be in error. The Chambers definition of evolution is "... the doctrine according to which higher forms of life have gradually arisen out of lower ... ." Moran is likewise judgmental of Webster, who defines evolution as "... the development of a species, organism, or organ from its original or primitive state to its present or specialized state ... ." If those definitions are not correct, then nobody but Dr. Moran knows what evolution is. Dr. Moran said this just after he quoted Dr. Douglas J. Futuyma as saying, "Biological evolution may be slight or substantial; it embraces everything from slight changes in the proportion of different alleles with a population (such as those determining blood types) to the successive alterations that led from the earliest protoorganism to snails, bees, giraffes, and dandelions." Excuse me, but isn't that what the definition in Webster's dictionary said evolution is, except that Webster is much easier to understand?

I strongly suspect that this smoke screen is sent up to conceal the truth. I'm thinking that Dr. Moran is the one who does not understand what evolution is. Evolution is the poison of society. Evolution is Satan's doctrine.

Psalm 24:1 declares that *"The earth is the the LORD's, and the fulness thereof; the world, and they that dwell therein."* The fullness of the earth refers to everything in or on or about the earth, even to the microscopic things in living cells that man can only gaze at in wonder if he understands that it is a created thing. If he gazes at it with the profane eyes of an unbeliever, then he will see only moving matter that has no significance in the real world of toil and weariness. Not for anything would I wish to devalue the knowledge and work of brilliant scientists, whatever may be his religious belief. However, I am determined to point out as best I can the error of their way: the error of their way being their refusal to give the Creator glory for His work of creation. Scientists are working with the products of God's labor. Whatever man sees or touches God was created by God from nothing and it belongs to Him. Scientists have a divine right to examine these things, for God said in the beginning that man is *to*

"... subdue the earth: and have dominion over the fish of the sea, and over the fowl of the air, and over every living thing that moveth upon the earth."

We would be much better off without some of the products of science -the internal combustion engine, for example -but most of science has helped us.

The non-renewable resources of the earth were created when God created the earth. Lead, iron, probably coal and oil, and all the other metals, both base and precious, were created in the earth in verse 1 of Genesis chapter 1. Brass and iron are both spoken of in Genesis 4:22, and steel is spoken of in II Samuel 22:35. Gold is spoken of in Genesis 2:11, and Ezra 8:27 speaks of copper. There are six metals spoken of in the Old Testament. There are thousands of other things in the earth's crust that are important to man, but all of them belong to God. We are only stewards of what God has endowed the earth with.

The great apostle used this verse in I Corinthians 10:26 when he was talking about what man is permitted to eat. I have heard people talk about it being unhealthy to eat pork, and how folks shouldn't eat other stuff, and I am not pushing against their ideas, when I say that I believe a person can eat just about anything in moderation, if he asks God's blessings upon it. I dearly love picnic shoulder. This eater is not a recreational eater. I eat for fuel, but I do enjoy eating. Picnic ham is one thing that I have difficulty with. I can eat it just for fun. Especially when it is surrounded by plump, luscious yams.

I think the main reason a person's diet is harmful to him, if it is, is that he doesn't take time to thank God for what is on the table, and the hands that provided it, and the hands that put it on the table. Slouching on a couch or easy chair watching tv is a great way to develop stomach problems. America is hauling around enough fat on her ribs and hips to share with everybody on Earth. *"The earth is the the LORD's ..."* and He has told us to eat of it, even if it were offered to an idol.

The LORD has even allowed some of us to call some of His Earth our very own. Isn't that wonderful? God has given my wife and me a speck of His Earth that's nearly five hundred feet along the street, and nearly one hundred fifty feet deep with a nice little house sitting right in the middle of it. I hope you have some of God's Earth to call your own. I hope if that is so, you are taking good care of it, because He will reclaim it someday.

I am happy that this verse in the Jewish hymn book makes it clear that I, a dog of a Gentile, also belong to the LORD. The psalm says, *"... the world, and they that dwell therein."* Oh, bless God! I belonged to Him while I was a sinner. I was His because He created me. I was one of His creatures in the world that He owns. And then I gave myself to Him. I acknowledged His ownership. I was happy to confess that I was His own creature. Blinded, vilethough I was, I was still one of His very own. Now I am a spiritual son of **Abraham**, as entitled to be called a son of God as any Hebrew upon the earth.

In turn, what a rebuke this is to those who say that peoples of other nations, kindreds, and tongues do not rightfully belong to the LORD. He created them. They belong to Him by creation. They are part of the world that God loved and sent His Son to redeem. God will dispose of his own in the way He chooses. And the criteria of His choice have been made. His choice will depend on how his own dispose of Him.

THE BREATH OF GOD

My prayer is that God will "Create in me a clean heart O God, and renew a right spirit within me." What a good place the old world would be if every one of us prayed that prayer, and then allowed the Holy Spirit to perform the work in us.

Psalm 33 is another psalm that has in it references to the creation work of the the LORD. Let's read verses 6 through 9, and rejoice in the rich Words.

"By the Word of the LORD were the Heavens made; and all the host of them by the breath of his mouth. He gathereth the waters of the sea together as an heap, he layeth up the depth in storehouses. Let all the earth fear the the LORD: let all the inhabitants of the world stand in awe of him. For he spake, and it was done; he commanded, and it stood fast."

When I was a lad of about seven or eight, I was doing skin-the-cat tricks on a tree limb when I lost my grip, and fell to the ground. I was upside down when I fell, and so I hit the ground flat of my back. All of the breath was knocked out of me, and I lay for a long time (at least it seemed like a long time) gasping for breath and thinking I might die. Eventually, my lungs were able to function again, and as breath came again, I lay there for awhile thinking about how wonderful it was to be able to breathe.

We are told that when the Saviour was impaled to the old rugged cross, that hanging there with his weight pulling down on his arms, that it would have been very difficult for him to breathe. We know that He made only seven brief statements while hanging there, and perhaps, the physical difficulty of breathing was a factor in limiting what He said.

We are told in Genesis that God breathed into Adam's nostrils the breath of life. How intimate that seems that Christ the Creator would get so close to the face of this clay image of Himself that He could breathe into his nose. Adam must have instantly started to live, and the first thing he saw was his Creator's face almost against his own.

A king does not sweat from labor. Jesus sweated when he was a carpenter, but He had divested Himself of His royal position at that time. A king gives commands. A king does not scream his commands, unless he has allowed his rage for some reason to rob him of his composure. He speaks, and his spoken Word carries as much authority as if he screamed or raged.

God needed not to raise His voice to bring matter into being. Matter did not exist before He commanded it to appear, but when He gave the command, matter, which did not exist, obeyed Him, and came into being. God's Words were with such power that the atoms were formed out of that energy. *"For he spake, and it was done; he commanded, and it stood fast."* Isn't it something to realize that everything obeys God, even stuff that didn't exist, except humans and the devil and his **angels**?

This great psalm bears no title or author's name, teaching us, according to Dickson, "... to look upon holy Scripture as altogether inspired of God, and not put price upon it for the writers thereof." A psalm of creation certainly needs no human author named in order for it to have authenticity. God and some of His holy angels were alone at creation, and no man could bear witness to the truth of creation. God reveals to us the method of creation -He spoke, and it was done. The Word "breath" here points us to the Holy Spirit. God in His triune Being was involved in all the major movements of history. God the Father, God the Son, and God the Holy Spirit were all involved in the work of creation, just as they were all involved in the salvation of man. It was Christ Who did the actual work of salvation, but the Father and the Holy Spirit were also involved. It was Christ Who did the actual work of creation, but the Father and the Holy Spirit were also involved. In Genesis 1:1, God uses His plural name "Elohim" to indicate that all three persons of the **Godhead** were involved. Scholars call this a "uni-plural" name.

In Hebrews 11:3,

"... the worlds were framed by the Word of God, so that things which are seen were not made of things which do appear..." When I saw the little wren on the back of the chair on my front porch this morning, I saw a thing that was not made from things which do appear. I saw a thing that was made from atoms which cannot be seen, and the atoms were made from the energy of God's Word, or breath.

If God's people could comprehend what it is we have in the Word of God, there would be no obstacle too great to move. Jesus comprehended it when He said,

"If ye have Faith as a grain of mustard seed, ye shall say unto this mountain, Remove hence to yonder place; and it shall remove; and nothing shall be impossible unto you."

Complete confidence in the creation story in Genesis will increase the Faith of the believer. Understanding as much as possible about what Christ did is a labor with tremendous benefit. Augustine said, "Let any make a world, and he shall be a God." Here is the manner in which to become a god. Create a world. Evolution and Mormonism both teach that if a man achieves certain things, he will attain unto godhood. Evolution must eventually result in man being gods, and Mormons believe that if a man is good Mormon, then he will attain to godhood. For a woman Mormon, she can only hope her husband makes it to the big time, for she goes along with her husband to whatever fate.

What a pity that people are still deceived by the lie that Satan told Eve in the garden of Eden. You would think that people with any gray matter at all between their ears, would understand that a man can never be a god. A reptile could evolve into a chicken quicker than a man could evolve into a god. God is the Creator. No one else but God could ever have such power. Faith in God the Creator and Redeemer brings a pleasant confidence in God that gives the soul complete assurance and rest.

"For he spake and it was done, he commanded, and it stood fast." His Word and His power are the same. His speaking and His doing are the same. He spoke and it was done. In Latin, it is Dictum factum. SAID DONE. If a child tarries thirty seconds before obeying, that child is disobedient for thirty seconds. The creation was not so. When God spoke it was done. Creation was not a thirty-second delay. When God spoke it was done.

Breath conveys all sorts of Words, and our Words convey all sorts of emotional meanings. When a husband puts his cheek against the cheek of his wife, and whispers, "I love you," she can feel the breath of his lips that conveyed the message. There is a softness about feeling someone's breath.

Breath both goes in and comes out. When breath goes in, it provides essential oxygen to the breather to sustain life. When breath goes in, it is harmless to others. When breath goes out, it can mean death or life to others, depending on who the breather is, and what his mood is. Lions kill by strangulation. They take their prey by the throat, and hold on until the creature succumbs from lack of oxygen.

The breath of God has always conveyed love and grace. Even when He has been forced to terminate the existence of certain people, His breath has been heavy with grace. To protect righteousness, evil must be attended to. If evil resides in a person, then that evil person must be dealt with for the protection of the good. God instituted governments to protect good people from evil ones. Grace will destroy the evil one to protect the good one. Sparing a murderer to kill again is not showing grace. Grace and love will kill the evil ones among the good in the interest of peace and safety.

When Words of hate or malice issue forth from the mouth, then the breath is not felt, for these Words are nearly always shouted or screamed from a distance. Being close means love, being distant means hate. Breath is felt when one is near.

Blessed be the old hymn that Edwin Hatch and B. B. McKinney wrote more than a hundred years ago, "Breathe on Me."

"Holy Spirit, breathe on me, Until my heart is clean; Let sunshine fill its inmost part, With not a cloud between.
Holy Spirit, breathe on me, Till I am all Thine own, Until my will is lost in Thine. To live for Thee alone. Breathe on me, breathe on me, Holy Spirit, breathe on me; Take Thou my heart, cleanse every part, Holy Spirit, breathe on me."

How comforting to feel His divine breath upon our cheek as we are told that we can

"... understand that the worlds were framed by the Word of God ..."

A CLEAN HEART

The Word "create" appears in the book of Psalms only one time, but the Word "created" appearsfour times. The Hebrew Word "bara" means create, but it is also used of the past tense in English. "Create" means "to set going, or to bring into existence." In that sense it is possible for man, or even beavers to create. Men create works of art or literature, and beavers create dams. In the strictest sense, however, create means to make something physical out of nothing. Matter is anything that has weight and takes up space. The creationof matter is something that only God can do.

Evolutionists who deny the existence of God, occasionally use the Word "create," but it seems obvious to anyone that matter cannot create anything. Evolutionists now use the Word "design" sometimes, but there can't be a design without a designer, but the evolutionist rejects a Designer.

Victorinus Strigelius back in the 16th century said that "This psalm is the brightest gem in the whole book, and contains instruction so large, and doctrine so precious that the tongue of **angels** could not do justice to the full development." This psalm is often called the sinner's guide. Athanasius recommended that all Saints repeat it when they awake at night. It is a psalm that I often quote in writing as a prayer, so that I might enjoy the power of a clean heart, and a humble spirit.

David wrote this psalm in which he cried, *"Create in me a clean heart, O God, and renew a right spirit within me."* **David** lived a thousand years before Calvary, and he had not the privilege of calling upon a crucified Saviour. He knew not about the new birth. But he did have a very keen sense of his sinfulness as a mortal man. In verse 1 of Psalm 51 he pleads for mercy, and speaks of his transgressions. In verse 2 he again speaks of his sin. In verse 3 he acknowledges his transgressions, and says his sin is ever before him. In verse 4 he again acknowledges his sin, and speaks of his evil. In verse 5 he laments that he was shapen in iniquity, and was conceived in sin. Before there can be cleaning, there must be an awareness of a need for cleaning. Nobody is going to set out to clean a clean house.

Then **David** turns to the grace of God in verse 6, and exalts God Who alone is able to make the vilest sinner clean. In verse 7 he calls upon God to purge him and wash him. In verse 9 he pleads for forgiveness, and in verse 10, he prays, *"Create in me a clean heart, O God, and renew a right spirit within me."* David is so convicted of his sins that he knows that his only hope is a new heart. Remodeling the old one will never do. It is too blackened with sin. It is too filled with iniquity.

Had **David** been living a thousand years ago instead of three thousand years ago, he could have called upon the the LORD for an application of the blood of the LORD Jesus Christ for cleansing. Today the old heart is left to decay and die, and a new heart is created when we are born again, and the new heart is kept clean by the blood of the Lamb. Psychologists and psychiatrists cannot understand the human psyche until they understand that man is a sinner who can have a new heart. People do not need to be counseled as much as they need to obey God's Word.

What did it require to create the physical heart that beats within our chests? An organ so complex and endurable is a marvel of design. It is incredible that a muscle can work for a hundred years and never rest, never miss a beat. The heart is probably the most endurable living organism on Earth. Trees live for thousands of years, but they are unconscious, and grow extremely slow. When God formed man of the dust of the earth, when He came to the heart, surely, He must have taken extra care in putting it together. The notion that the heart evolved without any guidance or intelligent planning by an intelligent, living being, is to assert that matter can create itself.

But **David** was calling upon God to create an even more complex and enduring thing. **David** was referring to his soul, his very inmost being. **David's** soul was soiled by his blood guiltiness. God paid an awful price to make it possible to create in us a clean heart. **David's** old heart was hopeless. It could not be repaired. So vile was it that it was beyond hope. **David**'s prayer today would be, "Forgive my sins, O God, and save me for Christ's sake." Our old nature is left untouched by the grace of God. God creates us a new man when we are born again. The old heart is left in place to make us sick, and to torment us all of our life, and to be the target of Satan's darts. Our old heart is left to cause us to doubt, and to fear. Our old heart is left to be redeemed at the resurrection.

Paul described his troubles caused by his old rotting heart in Romans 7:24.

"O wretched man that I am? Who shall deliver me from the body of this death?"

A body of death has a heart of death. A dead heart, stored in formaldehyde will never pump a drop of life-supporting blood. A dead soul will never raise to the Throne of Grace a Word in prayer. A dead heart, lying on a table beside an operating table, will never beat faster at the sight of a beautiful thing. A dead heart, keeping alive one who has mocked at sin, will never throb to the melodies of old hymns. A man must have a new heart, and it can be provided only by Him Whose voice startled the non-existent matter in Genesis 1:1, and caused it to come into existence.

David also craved a right spirit, and without a newly created heart, he was not able to have a right spirit. David had enemies who were determined to kill him. His enemies were chasing him to death. Yet, **David** wanted a right spirit. With a divinely created heart, David could have a right spirit. That **David** had a right spirit is demonstrated time after time in his interaction with his bitter enemy. **David** and his warriors saw Saul asleep, and his men wanted to go and kill Saul, and they could have done it, but **David** said "No." **David** mourned when he heard of the death of Saul. With a newly created heart, we, too, can live in a way that will make us a powerful witness to the grace of God. After God creates a new heart for us, we need be concerned only with keeping it clean. Yes, the created heart gets soiled. Our walk in the world every day soils our feet, and this is why Jesus said to Peter,

"... He that is washed needeth not save to wash his feet, but is clean every whit: and ye are clean, but not all."

Jesus knew that Judas did not have a created heart. We are prone to sin, and wander away from the God Who loved us and gave Himself for us. Further instruction is found in I John 1:9,

"If we confess our sins, he is Faithful and just to forgive us our sins, and to cleanse us from all unrighteousness."

We should glory and marvel in the grace of God more than anything else in human experience. The grace of God that goes out to every human offering forgiveness of sins when we don't deserve to be forgiven for sins. When we understand that God will forgive anyone, no matter what his crime may be, or how long he has lain in the crimson dye, God will forgive him if he will come to God in Faith believing and receiving. I am struck as I read Jeremiah how the grace of God is sent forth to His erring people over and over again. All of us have our various shortcomings and trespasses. I have mine, and you have yours, but God is the God Who will cleanse us from all sin. We can live with a clean heart if we live close to the gracious God Who is grace personified.

In Verse 12 of our chapter, **David** pleads, "Restore unto me the joy of thy salvation; and uphold me with thy free spirit." There is no cost in coming to Calvary. The joy of God's salvation is lost only when we forget it. Keep in mind the moment you were born again. Revel in that moment, and allow the joy of God's salvation to be a flame in even the darkest time in life. The joy of God's salvation comes with a clean heart. The joy of God's salvation has seen the open grave. It has been to the unemployment office, to the prison, to the nursing home. The joy of God's salvation will do us when we are dying.

"Create in me a clean heart, O God, and renew a right spirit within me." The God of creation will abundantly answer that prayer. God wants to perform that creative act, and only the resistance of the individual prevents the LORD from creating a new heart for the unbeliever.

WONDERFUL WORKS

Verse number 5 of Psalm 40 speaks of the wonderful works of God our Creator. Life has so much more meaning when we acknowledge the work of God in the universe. A personal experience with the LORD is deep and rich, and such experiences are ours for the asking. God is always available to bless our lives with His nearness and His grace. When I eat a fine ear of corn, I enjoy it infinitely more because I know that God created the matter the corn is composed of, and He set in motion the Laws of physics and chemistry that caused the corn to grow.

Psalm 40:5 says, *"Many O* LORD *my God, are thy wonderful Works which thou hast done, and thy thoughts which are to us-ward: they cannot be reckoned up in order unto thee: if I would declare and speak of them, they are more than can be numbered."*

Writing and preaching about the wonderful works of God is discouraging to me in some ways, because it seems that what I say about them is so mundane. How can I speak Words that can contribute to what the Word of God says? Even with a doctor degree on my wall, I stand beneath these great statements like a blubbering little boy trying to explain to his playmates just how great his daddy is. I feel just plain tongue tied and inadequate. But such a task the LORD has charged me with.

Every work that God has done is wonderful. There is not a single thing we can find in the Holy Writ that God has done that is simple or ordinary. God leaves the simple and the ordinary for us to do. For example, in the book of John where is recorded that time when Lazarus died. Jesus stood at the tomb of his friend, and wept. Mary and Martha believed that He could have prevented Lazarus' death, but they needed to take another step of Faith, and learn that

- 165 -

JesusHimself is the Resurrection and the Life, and that He Who had sustained Lazarus in life could deliver him to life from death. Jesus was ready to raise Lazarus from the dead, a task certainly, that no human could do. The healers on tv do not even make a pretense of raising the dead. The greatest men of God's choicest servants have not been able to raise a dead body, but nearly anyone could move a stone, with the help of some stout men. And that is what Jesus demanded. People must move the stones. In John 11:39 it is recorded that

"Jesus said unto her, Take ye away the stone. Martha, the sister of him that was dead, saith unto him, *LORD*, by this time he stinketh: for he hath been dead four days. Jesus saith unto her, Said I not unto thee, that, if thou wouldest believe, thou shouldest see the glory of God? Then they took away the stone from the place where the dead was laid. And Jesus lifted up his eyes, and said, Father, I thank thee that thou hast heard me. And I knew that thou hearest me always: but because of the people which stand by I said it, that they may believe that thou hast sent me. And when he thus had spoken, he cried with a loud voice, Lazarus, Come forth. And he that was dead came forth, bound hand and foot with graveclothes: and his face was bound about with a napkin. Jesus saith unto them, Loose him, and let him go."

Humans could move the stone. They could not raise the dead. God performs the works that man cannot do. That is why Jesus died on the cross in our stead. Only He could perform the wonderful work of man's salvation. No other man or god exists who could perform such a work.

God created animal and human life in such a way that it can adapt to its environment. There are squirrels living near both the north rim and the south rim of the Grand Canyon. The squirrels that live on the north rim look very different from those on the south rim, even though they are all squirrels. These squirrels never meet. It is about ten miles across the Grand Canyon, and the canyon is a mile deep. The squirrels can see across the canyon, but they are separated by a barrier that they cannot cross. They have adapted to their separate environments which are quite different.

Though these squirrels are different, they are still squirrels. Adaptation is not evolution. The squirrels adapted, but they did not evolve. Creatures must have a capability to change that they can get used to a new climate, food, surface conditions, and many other influences, but they remain whatever they were. They do not evolve.

I believe God must have created the animal kingdom for the wonderment and amazement of man. A thing as lowly as a frog can be a marvelous example of God's many wonderful works. In Australia in 1973 scientists found a frog which gives birth through its mouth. According to "Creation" magazine (vol. 15 no. 2, ppg. 26, 27) "The mother frog swallows her **eggs** after fertilization, and then stops feeding. For six or more weeks **eggs** develop and pass through a type of tadpole stage, all in the stomach, without being digested or passed forth along the intestine."[1]

The best story teller among evolution philosophers could not make up a myth that could explain such a phenomenon. Evidently, the **eggs** are covered with stuff that stops the digestive fluids from entering the stomach, and prevents the digestive tract from passing the **eggs** or the young tadpoles on down the digestive tract. In the stomach, the **eggs** hatch, and pass through a tadpole stage, as many as twenty-six of them, stretching the mother's stomach to a very thin membrane. When they are fully developed, she goes to the surface of the water, and allows them to step off her lip into a life of their own. *"Many, O LORD God, are thy wonderful works which thou hast done."*

It is a great pity that **evolutionist**s attribute God's wonderful works to natural selection and the other fabricated means of evolution. The lake in Tanganyika is teeming with beautiful and amazing fish. National Geographic in its special on public television had a story to attribute all these creatures to evolution. They were not creations of an omnipotent Creator, they were simply the result of matter creating itself, or moving itself into more complex forms. This is a pity for people who believe such fables, but it is most dangerous for the people who are responsible for this, because God will not permit people to handle Him and His Word so contemptuously.

A thing as lowly as a frog can be a marvelous example of God's many wonderful works. In Australia in 1973 scientists found a frog which gives birth through its mouth. According to Creation magazine (vol. 15, no. 2, ppg. 26, 27) "The mother frog swallows her **eggs** after fertilization, and then stops feeding. For six or more weeks

- 167 -

eggs develop and pass through a type of tadpole stage, all in the stomach, without being digested or passed forth along the intestine."

The best story teller among evolution philosophers could not make up a myth that could explain such a phenomenon. Evidently, the eggs are covered with stuff that stops the digestive fluids from entering the stomach, and prevents the digestive tract from passing the eggs or the young tadpoles on down the digestive tract. In the stomach, the eggs hatch, and pass through a tadpole stage, as many as twenty-six of them, stretching the mother's stomach to a very thin membrane. When they are fully developed, she goes to the surface of the water, and allows them to step off her lip into a life of their own. *"Many, O LORD God, are thy wonderful works which thou hast done."*

Salt, which is essential to life, is composed of two deadly poisons. Water, which puts out fires, is composed of two highly explosive elements: oxygen and hydrogen. Plants receive the energy of the sun, and convert it into matter. Given enough time, uranium decomposes into lead; potassium decomposes to argon gas. Isn't it amazing to realize that elements will decompose, but not one ever composes or becomes more complex. Evolution declares that everything is getting more complex. Gravity is so common that it is rare that anyone takes time to think of the wonder of it, and the mystery of it. The magnetic field of the earth is a wonder that man can only guess about. Consider the distant stars, and other objects in the vast universe.

My lips are numb when I endeavor to speak of the things that God has done, and is doing and will do, both in His creative work and in His redemptive work. Yet will I lift my voice in praise and worship to Him Who has created me and redeemed me.

THE EVERLASTING GOD

Please turn your attention with me to the 90th Psalm, and let's look at verses 1 and 2 ." LORD, thou hast been our dwelling place in all generations. Before the mountains were brought forth, or ever thou hadst formed the earth and the world, even from everlasting to everlasting, thou art God." The psalmist rejoices in the ageless God. This is the God that was before man existed upon the earth, and so He is not the figment of some fanatic's imagination.

This psalm is the oldest of the sacred hymns, having been written by Moses five hundred years before the other psalms. Moses, the great Lawgiver, wrote of the God Who brought forth the earth and its ancient mountains. Moses could look back to the dim past. He could look back past Noah. He could look back past Enoch all the way to Adam. He could look back past Adam to the empty space that only God inhabited. Moses knew that God had always been. He is the eternal God. There can be only one eternal God.

Since that day when Moses wrote his great psalm, 3600 years have passed, and we can look back to Moses, seeing that God has not changed one whit since Moses extolled His eternal virtues. We can look back past two thousand years of history to the life of God on Earth, and know that the prophecies that Moses wrote of Christ have come to pass. Jesus said, *"For had ye believed Moses, ye would have believed me: for he wrote of me."* We can look back past the prophets, and the writers of the other psalms to Moses, and even past Moses to see perhaps even more clearly than Moses that God is the everlasting God.

God has been the dwelling place of believers ever since God gave the breath of life. We are God's dwelling place, and He is our's. I Corinthians 6:19:

- 169 -

"What? know ye not that your body is the temple of the Holy Ghost which is in you, which ye have of God, and ye are not your own?" Acts 17:28, "For in him we live, and move, and have our being; as certain also of your own poets have said, For we are also his offspring."

The age of the earth is a mystery. The rocks will not reveal their secrets. God knows the age of rocks, and no one else. Evolution philosophers who are determined to convince the world that there is no God, insist that there is technology that will pry open the secrets of the rocks and reveal their age, but such is not the case. Creatopm scientists are unaware of a single instance where any artifact over 5000 years old has been tested, and the test result accepted by **evolutionist**s without dispute. When the rock that a fossil has been found in is tested, the age of that rock, and hence the age of the fossil, is not accepted unless that date matches the date pre assigned to the fossil. No date for a rock will be accepted by evolutionists unless that date matches the date they have already assigned to the fossil. Rocks are dated by the fossils in them.

Mr. Richard Leakey is a well-known fossil hunter and evolution philosopher. He is the son of Louis Leakey, another famous fossil hunter and evolution philosopher. Mr. Leakey works in the Olduvai Gorge in Africa. In 1972 he found a skull at Lake Rudolf which was to become famous. It was named the "1470" skull. The skull was " ... fractured into a great many small pieces, but when painstakingly reassembled looked very human indeed. The specimen was quite unusual. It consisted of a virtually complete skull except for the lower jaw, but opinions regarding its position in the lineage of man were sharply divided."

What is of immediate interest to us here is the story of how the date for the skull was derived. Potassium-argon testing gave the fossil an age of 2.6 million years. That age would greatly upset the evolutionary time scale, because that would make the skull older than any of the missing links. Most evolution philosophers decided that the date was wrong because they could not afford to have the time scale upset. "The first sample of KB teff, the rock in which the 1470 fossil was embedded, gave an average age of 221 million years (Fetch and Miller 1976). This was about one hundred times greater than expected, so another sample was sen the laboratory, and an age of 2.6 million years, plus or minus 260,000 years, wasproduced and found acceptable." That made 1470 the oldest man, and that is what the headlines proclaimed to the world in spite of the extremely questionable evidence. More tests were made on the rock with dates from 290.000 to 19.5 million years. After a great deal of controversy among evolution philosophers, the age of 1470 was quietly reduced to 1.8 million years

Testing a piece of rock picked up in the field, or hammered out of a cliff, is preposterous. No other science would think of testing something that had even one unknown about it. A chunk of rock from the field is totally unreliable as a laboratory specimen.

"Before the mountains were brought forth, or ever thou hadst formed the earth and the world, even from everlasting to everlasting, thou art God."

How beautiful are the Words. How beautiful is the thought. The unchanging God. The eternal God. How could God be otherwise?

If anything personifies age, it is the mountains. When I was in the fourth or fifth grade, my teachers taught us that the Rocky Mountains were new and sharp, but the Appalachians were old worn down mountains. I loved the gorgeous mountains that surrounded my home in east Tennessee. I used to hike those mountains often. I have slept many nights under the stars on the top of Holston Mountain, and others. I have sat on the top of great mountains. and gazed into the distance where I could see other mountain ranges. It is not hard to understand why Indians could think of mountains as being holy places. The higher one gets to Heaven, the closer he would be to God, they reasoned.

Before there was a mountain, or an earth to sit one on, God was God. This is my God. My God never changes. Those majestic mountains I loved so dearly change from day to day. Wind and rain carry soil and rocks off their slopes, and float them into the Atlantic Ocean or the Gulf of Mexico.Earthquakes can change mountains. Forces in the earth beneath mountains can raise them toward the sky. But my God, Moses' God, **David**'s God, Adam's God can never change or be changed. Darwin and his followers have changed. Some are dead. Others are dying. They change their minds about what they believe from time to time. Their literature changes. It seems they know nothing of eternal values. My God never changes.

He needs not to change. How could perfection change? How could perfection cease? God is forevermore.

Because we are undergirded with the knowledge that God is eternal, we need fear nothing. That is why II Timothy 1:7 is such an eternal truth.

"For God hath not given us the spirit of fear; but of power, and of love, and of a sound mind."

Why should I fear anything? My God is the same God Paul worshiped, served, and trusted, and died for. He has not changed nor gotten older in the interim. He is not senile or suffering from memory loss.

Faith can be very timorous stuff. Flesh was created to interact with physical things like rocks and trees and water. The five senses give assurance that we are in a certain place with a certain environment. We feel, see, smell, touch, and taste things around us that give us confidence, but even these senses can often fail us. The soul and spirit have no such senses. The Rock that Faith is anchored to we have never experienced with our physical senses. Still, Faith can give us more assurance than our physical senses. We can know certain things by Faith that we cannot know by knowledge or intellect.

Faith does not prevent one from experiencing fear once in a while. Fear can be a life saving thing. Fear of a rattlesnake will prevent us from getting too close. Fear of electricity will prevent us from touching bare wires. A *"spirit of fear "* is a fear of the unknown that torments. Faith takes that away.

In verse 10 of our beautiful psalm, Moses says, *"The days of our years are threescore years and ten; and if by reason of strength they be fourscore years, yet is their strength labour and sorrow; for it is soon cut off, and we fly way."* Shortly, I shall enter into it fully, and why should I be afraid? A stone idol will someday dissolve. A wooden image would disappear even faster if its owner got cold one winter night. A god in the mind will change or pass away. God is everlasting.

STORMY SEAS

A gain the psalms speak of stormy seas in 107:23. The only place in the world that calm seas can be found is in the heart of the one who is resting in the sacrificial work of Jesus Christ. Why should it be a matter of perplexity that men are filled with sin, and therefore are like the rolling seas that cast up mire and dirt. The message of peace in Christ is more common today than it has ever been, and is being ignored by the same multitude.

Psalm 107 is another of the psalms that speak of creation - what the LORD has made. This time it speaks of the sea. The Word "ocean" is not in the Bible (It is found twice in the NIV, but that is not the Bible. It is a man-made book that has at least one heresy in it [Luke 2:22]).

"They that go down to the sea in ships, that do business in great waters; These see the works of the LORD, and his wonders in the deep."

Most of us have seen drawings on very old maritime maps of dragon-like creatures attacking ships. There is also a host of stories told by seamen of old about great monsters in the sea. Man has a natural fear of the unknown, and we can assume that some of these old tales arise out of men's fear of the unknown sea, and their efforts to explain things they did not understand.

On the other hand, there is always an element of truth in myths and old tales, and so it may be that there were creatures in the sea that are no longer there. Ships in ancient days were small, and men would naturally have a greater fear of creatures that people on board big modem ships would pay little attention to. Whatever they saw would also cause them to wonder at such things. They would have seen nothing on land that resembled those great creatures seen in the ocean. In 1977 a Japanese fishing vessel caught the carcass of a great sea animal in its net. The marine biologist on board made a study of the creature. He made pictures and took tissue samples, and then, because of the fear of contaminating their fish cargo with the body of this unknown creature, they put the carcass back into the ocean. Consequently, Japanese scientists made a thorough study of all the data brought to them, and concluded that the animal was a sea-dwelling dinosaur called Plesiosaur. It was 30 feet long.

We have no idea what might be yet in the oceans of the world, since only about 10% of them have been explored. But we do not need to be scientists to see the work of God in the sea. Such things as we can see just by wading along the beach are enough to convince us that the sea teems with animals of enormous complexity. A creature must be whole and complete, fully suited to its environment in order to survive. The little fish that crawl on land for short periods of time are fully suited to that sort of behavior. If they remain out of the water too long, they dehydrate and die. An animal can only survive if it can live in its environment, and any change in its physical makeup would cause it to die, unless changes in its environment forces it to adapt.

Natural selection does operate in nature. Natural selection is the darling of **evolutionist**s, but it does not work to evolve organisms to a higher state of complexity. Natural selection is a conservative behavior. The strongest of a kind reproduce. The weaker members of the group do not reproduce. This maintains the purity and strength of the genes, and helps the kind to survive. Natural selection does not cause new kinds to appear in nature. Variations do appear when a kind is used by man, and its reproduction isregulated by man. The genes of a creature allow for a wide range of differences, but the genes forbid the formation of something different. Reptiles could never change into birds, no matter how many years they are given. There are simply not enough changes for the known number of atoms to make to bring about such changes.

If God had not created each creature to fill a specific niche in the system of living things, and if creatures could be constantly changing from one thing to another, there could be no stability on the planet.

The phenomenon known as symbiosis is an amazing behavior that enables two organisms to live together for the mutual benefit of the two. If you have seen many films on life in the seas, you have doubtless seen big predator fish having their teeth or scales cleaned by certain small fish. The predator will not eat the small cleaner fish. Crocodiles lay with their mouths open for certain birds to clean their teeth. The crock never eats those birds. How could creatures evolve together so that they could fit together like this? Some of these creatures cannot survive without the other.

I am more and more amazed at the dismal ignorance of many people as I search the internet. There seems to be a general lack of having ever read anything, especially the Bible. If people read the Bible as much as they talk about it, it would revolutionize the world. Let me give you another example of a thing written on the internet. This man says, "I think your idea of evolution is quite cloudy. If you were gods [sic] creation then why can't you survive without heavy clothes in the winter? You would die otherwise...I thought god [sic] created things that were perfect, why do you need s and other so called necessities. The reason why we can exist in winter is because we have developed ways to avoid winter. Most plants only live for a year, they are on a different time schedule than you." (I copied this as it was written.)

How could people be so ignorant and still operate a computer well enough to answer a note on the internet? This man does not spell God's name with a capital letter. If the person he was answering had a cloudy idea about evolution, it is certain that this person will never be able to explain it to him. I wrote an editorial to our newspaper, which I talked about in one of my books, and a man wrote a letter rebutting what I had said. His letter was so full of errors that I wrote him a personal letter suggesting that he do some reading and research in the literature before he got into something again that he didn't know anything about. Evolution high priests can say almost anything and people will accept it without question, while a preacher of the Word of God is questioned and doubted until the situation seems impossible sometimes.

I am aware of how poor I am as a teacher. I am painfully aware of my poor voice, and all the rest. But it seems to me that what I have to say is of such importance that people would be willing to suffer some discomfort to receive the message. I am praying that God will raise up someone who can take this ministry, and do a much more effective job of presenting its message than I can. This message must be heard. It is absolutely essential that people know that there are people who know the truth about the origin of the universe, and its end. If evolution is true, then the Bible is false. If evolution is true then there is no Gospel for there would be no need for a Gospel. Man would return to the dust and leave the mystery of life unsolved for eternity. People like the man I just quoted need help very badly, and I write letters on the internet to explain some of these things, but an ignorant person is not just an uneducated person. Most ignorant people refuse to acknowledge that there are things that they do not know, and will not listen to instruction.

Today men are going down to the sea in ships that can convey them down into the black depths of the ocean. Miles down they are finding wonders of the deep put there by the LORD Creator. I pray that some of those people will come to their senses and realize that what they are seeing could not have become what they are by blind chance. The mind-boggling diversity of these creatures should make it impossible for anyone to believe that they just happened to be there.

Jesus was acquainted with the sea. He sent Peter to catch a fish with a coin in its mouth.

"Notwithstanding, lest we should offend them, go thou to the sea, and cast an hook, and take up the fish that first cometh up; and when thou hast opened his mouth, thou shalt find a piece of money: that take, and give unto them for me and thee." (Matthew 17:27).

He told the fishermen where the fish were: *"Now when he had left speaking, he said unto Simon, Launch out into the deep, and let down your nets for a draught."* (Luke 5:4). He also commanded the sea to be quiet:

"And he arose, and rebuked the wind, and said unto the sea, Peace, be still. And the wind ceased, and there was a great calm."

He stood on the sea, and beckoned to me. I feared the great wave, but he made me brave, so I walked on the deep like the others of His sheep.

MAN'S EARTH

S ummer is glorious. Just like fall and winter and spring. God's glory is seen in every season. The humming birds are doing aerial combat around the feeders on the front porch and the patio. They sometimes sit on the perch above the feeder outside my study window, and preen. They must have a lot of things on them. Just about every living thing has some sort of parasite that feeds on it, even fleas, so the beauty of the humming bird just camouflages his mites. God's world is beautiful, even after 6000 years under the **curse**. The soil in my garden is richand good, but it is full of weed seeds. Thank God for the hope of His soon appearing when He will take us to our new home. This old home will then purified by fire. The earth may be drawn close enough to the sun to burn it. We can only speculate.

We continue to study the psalms that refer to the LORD's creative work, and this study has not been exhaustive by any means. Today we give attention to Psalm 115, verse 16. "The Heaven, even the Heavens, are the LORD's: but the earth hath he given to the children of men." This is an important bit of information to reinforce what we learned in Genesis 3:15.

"And the LORD God took the man, and put him into the garden of Eden to dress it and to keep it."

That command was never changed, and remains in force till this very day. Man is charged with the responsibility of keeping the earth and caring for it. It is right to be an environmentalist, if the right perspective is placed on all elements involved in the environment. We must remember that God created man to have dominion over the earth and its creatures.

God reserves the Heavens for Himself. We will one day have the privilege of walking in God's Heaven, and it will be our final home, but the Heaven of heavens will always be the private property of the LORD God. God needed no Heaven before He created the earth. He needed no abiding place, for the universe was His domain. But when He created Earth and its creatures, then He needed a place for His throne from which to rule His creation. He also needed a place where His chosen ones could live with Him forevermore. We know very little about Heaven. It is a material place as opposed to a spiritual place, though spiritual beings dwell there. It has streets of gold, and a river with trees lining its banks (Revelation 22:1,2). Jerusalem can be measured (Revelation 21:16), and so it must be a material place. I suspect that Heaven is the nucleus of the universe, and if that is true, then it must be unimaginably far away. Traveling at the speed of light it would take a space ship thousands of years to reach some of the stars in heaven, so how could Jesus have come to Earth to be born a man, and how could our souls get to Heaven in time to come back to Earth with the Saviour?

Those questions are simply answered and understood when we consider that our souls and spirits will not be bound by time, space, and gravity when they are released from this robe of flesh in which they are imprisoned. And, at the resurrection, our new body will be like the resurrected body of Jesus Christ, and will likewise not be confined to those three dimensions. We will be able to get to Heaven at the speed of thought. We can think of a place, and our soul will be borne to that place at the speed of thought. That is instantaneous.

Verse 15 of our psalm says, *"Ye are blessed of the LORD which made Heaven and earth."* It is almost too much for me to realize that I am a recipient of the blessings of the Creator. This is no idol we are talking about. This is the God Who inhabits eternity. I am poor and needy, yet I know that through the years my life has been before the LORD, and He has done things for me that I am not even aware of. He saved me when I was but a lad, and has watched every minute of my life. If we could somehow get this truth planted before us, how much better would be our life. God created the universe for our benefit and blessing. How strange. Here I am a living man holding in my hand soil that God created before He created Adam.

Many a war has been fought over ownership of the earth. Many a profane idol has been made from the material of the earth. God gave it to man. God reserved to Himself Heaven, and so He is able at any time to reclaim Earth for the benefit of those who love Him. How bad to abuse the earth. The worst invention ever made is the gasoline engine. Man misuses everything. Man never thinks about what he will leave for future generation. The ugliness of man has harmed the earth even more than we know, but God will make it right again. Jesus said, "Blessed are the meek: for they shall inherit the earth." (Matthew 5:5). Psalm 39:9 says, *"For evildoers shall be cut off: but those that wait upon the LORD, they shall inherit the earth."*

The great Joseph Parker wrote, "Dare the wicked man to read the Psalms? Has he any one of the hundred and fifty which he can call his own, and which he can read in the morning light before going out to renew his iniquity? Is there not one line left for the poor wretch? Has he not one string in all the infinite harps? Can he not quote one verse, saying, This encourages me to do the best I can for myself to perpetrate mischief, to outwit my fellow-creatures, to keep false weights and measures; this will enable me to give license to every desire of my heart? In all the Book of Psalms not one little line can be claimed by the bad man."

No wonder the bad man hates the Word of God. In it he can find not one Word of consolation or encouragement for his sins. Never can he find an honest excuse for his iniquity. Truly, the LORD gave the earth to the children of men, but He did not give them the right to shove other people off of it by hook or crook. Man must give an account for the conduct of his life, and of how he has served as a steward of the earth that the good God gave him for his happiness and peace.

God in His grace, created the earth as a warm, fruitful place for man to live in. How long man was to live upon the earth, we do not know. Perhaps forever, as God had provided the tree of life for his health. Perhaps God planned for man to live on Earth until the earth began to be crowded with people, and then God would take him to Heaven where there would be room for an infinite number of people. As a sinner living in the **curse**d earth, he lived nearly a thousand years, so we can assume that God intended him to live a long time on Earth. After Adam's sin, we know what God intends. It is His will that we live in holy pleasure on the earth and worship Him, respect the earth; be saved by faith; produce as many children as possible; lead all his children to the LORD; die in a ripe old age of 75 or 80 years. Ah, if mankind only had good sense...

Take note of Psalm 90:4: "For a thousand years in thy sight [are but] as yesterday when it is past, and [as] a watch in the night." What a beautiful statement. I'm glad God used a thousand years here. I wonder why people do not use this verse as well as II Peter 3:8, or instead of II Peter 3:8. When Peter wrote his statement, more than a thousand years had passed since Moses wrote his. The seventeen hundred years that had passed since Moses, had been no more than a watch in the night. Moses and Peter both knew that a day was never more than twenty-four hours. They would wrinkletheir brow and try to figure out how some men could make aday in Genesis mean a period of millions of years. They would wonder why men would want to do that. They never heard a silly idea like evolution that needs endless ages of time to change a lizard into a chicken. My God lived before the clock of time began. He lived before a mountain raised its ancient peak. God was never born, and He shall never die. He exists eternally, while we live everlastingly. God's eternal life had no beginning, and will never end. Out everlasting life began the day we were saved, and shall never end.

A SURE FOUNDATION

This book was being edited on November 8 almost two months after the cowardly, dastardly bombing of the World Trade Center in New York. In all the history of mankind, it would be difficult to find a more heinous crime. Such behavior was not motivated by God, but the devil, and Allah is his name.

I watched television as the huge towers burned, and I watched as they fell, and I wept. The heat generated by the air liners' full tanks of fuel weakened the steel framework of the buildings so that they could not support the enormous weight of the structures. The top floors acted like a great hammer coming down on the lower floors. The whole mass was ground into bits and pieces and dust. Human bodies were so crushed and pulverized that they could not be identified.

Contemplating the work of constructing sky scrapers causes one to marvel at the ability of man to raise such edifices. It is breathtaking to look down from the security of the top of such a building, but terrifying to think of standing on no more than a narrow beam hundreds of feet above the ground, with nothing to prevent a plunge to the earth below. It would indeed be fearful construction.

When we consider the construction of our body, we must stand in awe and fear of such a Creator Who is capable of building such a thing. Not only was He able to construct bone, blood, and flesh, but He made it all live. We studied this psalm only about three lessons ago, but we must touch it again for it is wondrous to behold. The living Word can never be exhausted, and it can never cease to bring forth fruit. We are more blessed than our limited minds can ever know in this world. God has given us good things in giving us His precious Word.

- 181 -

Psalm 139, verse 14 says, *"I will praise thee; for I am fearfully and wonderfully made: marvellous are thy works, and that my soul knoweth right well."*

Praising God and thanking Him for our wonderful body is a mark of nobility. To acknowledge God as the Designer and Creator of our body is to better understand our body. A physician who has acknowledged that God is the Giver of life and the Sustainer of life is one who can much more effectively treat us, and bring about improvements in our health.

A surgeon who understands that the body he operates on was created by the holy hands of the Creator, will more skillfully and tenderly touch that body. That doctor will not steal insurance checks that do not belong to him, nor will he steal from the government, meaning the people of the United States.

"I am fearfully made. . . " . I would be afraid to proclaim that I arose from a one-celled organism, and evolved through many stages of animals until I became me. I would be afraid that such impudence would cause the Creator to slap me. I would surely deserve it. What absolute and inexcusable nonsense to believe that the human body, the most complex system in the universe, could have developed by random chance. How irreverent can you get?

Dr. Morris wrote, "The absurd notion that such a marvelous organism could have developed slowly over the ages by random processes of evolution is a graphic commentary on man's desire to escape from God at all costs." It must be true that the motivation behind evolution dogma is merely an effort to eliminate God from the human consciousness. Little can be gained by studying man's origin, except satisfying curiosity. Since it should be obvious by now that there is no way to scientifically determine how life began, there is little hope that humanity could be helped or improved by the further expenditure of money on it.

The situation could be greatly improved if **evolutionist**s would simply stop talking about the earth being millions of years old. I watch as many programs as possible on science, and I cannot see how that talking about evolution and the age of the earth enhances these programs. They could give out just as much science by leaving out all these unscientific, philosophical references. There is simply no excuse for it.

Another area that would help greatly would be for **evolutionists** who are in control of the educational system to drop their prejudices against Saints and admit student Saints to science courses in universities. This is actually the most shocking and inexcusable prejudice practiced in America. A student who admits to being a creationist has virtually no chance to be admitted to a science degree program in this nation, and a professor who lets it be known that he is a creationist is in serious danger of losing his job, and will almost certainly be shunned by his colleagues, and cut off from the possibility of having his writing published.

These situations can only exist because evolution is a religious belief that makes fanatics out of otherwise reasonable and educated people. Evolutionists call creation science a "false science" or a "pseudo-science." Can there be such a thing as false science? If a scientist took as his life work the task of proving that the law of gravity is flawed, would he be labeled a false scientist? If a scientist took it upon himself to gather evidence to show that there are mosquitoes on the sun, would that be false science? It seems to me that any hypothesis should be worthy of scientific investigation if there is someone willing to undertake it. But in this, as in all other areas, liberals are notorious for standing on freedom of speech and everything else, except for people who disagree with them.

I am amazed by my own body. It continues to live and function after eighty-two years even though it has been sawed open and the heart cut open and an artificial valve installed. The appendix has been cut out. The bottom of the spinal column has been cut off. Bone has been chiseled out of its nasal passage. Half of its teeth have been pulled out. Three or four holes were punched in it and the gall bladder suctioned out. Skin cancers have been whacked off of it. Its blood sugar once went above a fatal level. It endured the insertion of about eighty radiated "seeds," and resulting agony for six months to kill cancer cells. A large section of the colon has been removed, and it has been saturated with chemicals for six months to kill cancer cells (nearly killling me in the process). This body has to be sustained by blood thinners and blood pressure pills and insulin shots. Yet the thing has taken that lickin' and keeps on tickin'. How can this be? All of us actually live one breath from death. I get no pleasure from thinking that my body is the product of natural selection that brought me from an apelike creature to a man who can walk upright. I can get down on my knees and pray. An ape can't do that.

"I am ... wonderfully made ... " Wonderfully made because I was made by a wonderful Creator. It's been a long time since I truly stood in wonder of something. Little children enjoy wonder more than anybody. When they see something unusual for the first time, their little eyes shine with wonder, as they consider what it is. Wonder is a mixture of curiosity, excitement, joy, and fear.

We often think of the wonderful statement of our LORD in Matthew 10:30 where He said, *"But the very hairs of your head are all numbered."* I love to pretend like I jerk a hair out of a child's head at conferences to demonstrate this. This is such happy thought, let's look at another verse. Verse 16 of our Psalm 139 says, *"Thine eyes did see my substance, yet being unperfect; and in thy book all my members were written, when as yet there were none of them."* This is simply marvelous! God made and saw every part of my body, while it was being formed in my mother.

Construction is a complicated business. Small things like models of buildings and bridges require much time, thought, and planning. People who build these things must have education and experience. There has never been a thing built without a brain guiding the builder. Order cannot come out of chaos. Explosions do not create anything. Construction requires attention to the smallest detail in the structure, whatever it is. A part small enough to lie in the palm of your hand can bring down a great building. It is amazing that a great building can have brick walls that stand hundreds of feet high. How do the bricks on the bottom layer stand the weight of the wall above them. It seems to me like they would be crushed. Did God have a set of blue prints before He began creating and making the cosmos? Is that what He did for the eternity before He began creating the universe – making blue prints? Would the planning for the universe, and all its components require God to think them all through? Each creature on Earth has parts that are complicated beyond anything we can imagine such as the eye of all creatures.

Introduction to Book IV

I enjoyed my trip on the Autobahn back in the forties when I drove an Air Force truck from Hanover, Germany to Oberpaffenhoffen near Munich. As we went out of Munich, we ran into a vicious storm while we were going up a very steep mountain. "Autobahn" was an early thirties Word. If someone had given me a book that was supposed to be two hundred years old, and that book had "Autobahn" in it, I would know that the book was not two hundred years old.

If I found a book that seemed old which had the Word "keyboard" in it, I'd know that the book was not old, after all. If you were digging in an archaeology site, and found a tv control, you would know at once that the site was not more than ten or fifteen years old, or somebody had gotten into the site, and planted the thing there.

The New Testamemt is the plant that the Old Testament root supports. Without the Old Testament, we would be left wondering what the purpose of the New Testament is. The creative work of God is an essential in the knowledge of man. The work of creation is what makes God unique. All of the gods of unbelievers are impotent: they do no more for man than any other inanimate object. A man would be as well served by bending his knees before an uncarved stone out of the wilderness as an ornate stone emblazoned with jewels. The New Testament is sprinkled with references to Christ's work of creation to emphasize His uniqueness.

The texts we have chosen for these essays are not all of the references to Creation in the New Teatament. My joy in the Lord was greatly enhanced when I learned that the truth of Creation is found scattered all through thre sacred text. I hope you will be so affected.

Psalm 89:12

"The north and the south thou hast created them: Tabor and Hermon shall rejoice in thy name."

Psalm 102:18

"This shall be written for the generation to come: and the people which shall be created shall praise the LORD."

Psalm 104:30

"Thou sendest forth thy spirit, they are created: and thou renewest the face of the earth."

Psalm 148:5

"Let them praise the name of the LORD: for he commanded, and they were created."

Psalm 51:10

"Create in me a clean heart, O God; and renew a right spirit within me."

Psalm 51:10

Book IV

MATTHEW AND GENESIS

With this essay I will begin a study of passages in the New Testament which refer to the first three chapters of Genesis. There are at least 63 references in the New Testament to the first three chapters of Genesis. This study is important because the Bible is a unity, and what is taught in one place in the Scriptures must be taught in the other. There is no such thing as a contradiction in the Word of God, and what one believes about a passage or doctrine in the Old Testament must be consistent with what the rest of the Bible teaches.

As a Bible literalist I use only the 1611 King James Bible, the A.V. of 1769, and I study it in the English language, meaning that I do not turn to the Greek or Hebrew meanings of the Words until I have gotten everything there is to understand from the simple English. That is not to say that I do not use definitions of the Greek or Hebrew Words at all. Word studies are important. I have found, however, that the English definitions are usually altogether sufficient for our understanding. God provided that Bible for ordinary people, and with the help of the Holy Spirit, any person can understand from the King James all he needs to know to live his life righteously.

We are not told in the Bible that Moses wrote the book of Genesis, but all reliable Bible scholars believe that he did. We are told in the Bible – Jesus told us - that Moses did write the Law, but Genesis does not include any of the Law of Moses (John 24:44). The events in Genesis 1-3 occurred centuries before Moses was born. That doesn't mean that he could not have written Genesis by divine revelation. *"All scripture is given buy inspiration of God, ..."* (IITimothy 3:16). He wrote the last chapter of Deuteronomy which includes the account of his death and burial, and he wrote that by divine revelation.

The point here is that Abel doubtless wrote the first four or five chapters of Genesis, and that these books were handed down to Methuselah, and from Methuselah to Noah, and from Noah to **Abraham**, and so on down to Moses. None of this is mentioned in the Word of God. However, in Genesis 5:1, we are told,

"This is the book of the generations of Adam. In the day that God created man, in the likeness of God made he him."

This book was probably written by Seth.

To think clearly about the days of Adam and the generations before the flood, we must free our heads of the cave man pictures instilled in our brains by our evolutionary public school education. The people who inhabited the earth before the flood were not grunting half-man-half-ape brutes. There is absolutely no scientific or biblical reason to believe such tommyrot. Men did live in caves in the past just as some do today, as **Job** described them in chapter 30. Adam and Eve may have lived in a cave until they could get tools made and houses built, but the people on Earth before the flood were the most intelligent peoplew ho have ever lived on the earth. Even today, we note that the most intelligent people can be the most corrupt. Adam was not a stoop shouldered hairless monkey.

"So God created man in his own image, in the image of God created he him; male and female created he them." (Genesis 1:27)

We place too much emphasis on science today, but we have been forced to turn to science to help counter evolution propaganda.

It would be inconceivable to think that Adam did not write books. A man of his intelligence and experience would be bound to make records of what he experienced. These books would explain several very important and interesting things in the Old Testament which I have already discussed. Doubtless, Moses had the books of Adam when he wrote the first chapters of Genesis, and he may have copied those books Word for Word under the Lord's direction. The Law of Moses is contained in the four books following the book of Genesis. But notice that God said

"Because that Abraham obeyed my voice, and kept my charge, my commandments, my statutes, and my Laws."

Abraham obeyed God's charge, commandments, statutes, and Laws 400 years before God gave Moses the Law on Sinai.

Where did **Abraham** get God's Law? I believe it was handed down to him from Abel.

The Bible is a unity. If Abel wrote the Words of Genesis 1-3, he wrote them by divine inspiration from the Holy Spirit of God. When Jesus said,

"...Have ye not read, that he which made them at the beginning made them male and female, And said, For this cause shall a man leave father and mother, and shall cleave to his wife: and they twain shall be one flesh?" (Matthew 19:4,5),

He was placing His divine guarantee on those Words that they were the Word of God. In recording them, Matthew placed God's divine approval upon what was said in Genesis 1:27; 2:23,24.

Incidentally, isn't that an embarrassment to note that Jesus asked if they had read? There are more Words in existence today available to man than ever in the history, and people read very little. Those verses in Genesis say,

"So God created man in his own image, in the image of God created he him; male and female created he them. And Adam said, This is now bone of my bones, and flesh of my flesh: she shall be called woman, because she was taken out of man. Therefore shall a man leave his father and mother, and shall cleave unto his wife: and they shall be one flesh."

We can now understand why a man who had no father or mother, that is, why Adam would say that a man should leave his father and mother for his wife. He was prophesying what would be the universal law for marriage. Both Jesus the Son of God, and Matthew, a divinely inspired writer of Scripture confirmed that what was written in Genesis was the divinely inspired Word of God, whether Adam wrote it, or whether Moses wrote it.

Many animals mate for life. Somehow we seem to have a special esteem for animals that mate for life. It seems so right, even for animals, to come together as a pair loyal and faithful to each other for life. There is a sense of maturity about being mated for life. There is an unselfishness about being mated for life. To be mated for life is so important that God said a man must leave his father and mother, to cleave to his wife. It is seldom a problem about a man leaving his parents,

This is the second commandment God gave to man. The first was that he leave the tree of the knowledge of good and evil alone.

- 189 -

This showed that the first thing man must do is to obey God, and allow God to have His place in the universe. God created man, but man should not think it pernicious, as most men do, to remain submissive to God. Most men think they are as good as God, and should at least share His throne.

The second commandment then was for a man to cleave to his wife. The reason for this is simple. God had punished the woman for disobeying Him, anddeceiving her husband. He had said,

"... I will greatly multiply thy sorrow and thy conception; in sorrow thou shalt bring forth children; and thy desire shall be to thy husband, and he shall rule over thee."

There would be times when the woman would be helpless in the child bearing work. She would need a Faithful and protective husband, and she would deserve his love and care during that time especially. God also made her desire be to her husband, and so it would be cruelty beyond measure for the husband to be at liberty to leave her, or desert her as he pleased. A man who does not care for his family has denied the Faith, and is worst than an infidel. There is no excuse for a man to abuse his wife. The laziest, sorriest mortal who ever lived shouldn't feel her husband's knuckles.

Of course, we see women whose desire is not to their husbands, and husbands who do not remain loyal to their wives, and that is called sin. It is also unnatural behavior - perversion. People who bewail all the heartache and misery in the world look everywhere for the answers except into the Word of God where the answers clearly are clearly found. God designed human interaction for the happiness and well being of the race, but man scorns the wisdom of God, and considers the wisdom of the world of much greater worth. When disasters occur, it is not the local pastors who are called in to counsel the children, it is the psychologists.

Man cannot improve upon what God has done and planned. Psychology and birth control pills and women's lib will never make for a happier life than simple, humble obedience to the Lord Creator Who loved us and gave Himself for us. God's ways are not grievous.

MALE AND FEMALE

G od gave mankind a pleasure he gave to no other creature when He allowed man to enjoy conjugal delights. All humans enjoy a special male-female relationship that the animals do not know. Courtship and marriage are uniquely human: introduced in Genesis, and unfolded in the New Testament.

Every book in the New Testament contains a reference to Genesis except Philemon, II John, and III John. There are at least 63 references in the New Testament to the first three chapters of Genesis. All of the New Testament books contain a reference to the first three chapters of Genesis except for Philippians, I Thessalonians, Philemon, James, I Peter, I and II John, and Jude. Of the books that have references to the first three chapters, there is an average of three references in each of the first three chapters of Genesis. The New Testament, like the entire Bible, is based on the first eleven chapters of Genesis. I want to express my gratitude to Dr. Henry Morris for providing the research for that information, recorded in his great book, *The Genesis Record.*

As with everything else a Saint believes, what he believes about the origin of the universe must be consistent with everything the entire Bible teaches.No verse in the Bible stands alone. Every verse is consistent with, and supported by, all the other verses in the Bible. The Bible is a unity. When a verse of Scripture, or a group of verses, is isolated, and used as proof texts for some doctrine, those verses must not be made to say something that would contradict in any way what the rest of the Bible teaches. Verses pertaining to baptism are a good example of such a practice. Error is the result of using parts of Scripture to prove something. We must begin with Scripture to form ideas and doctrines, rather than forming ideas and doctrine, and then trying to use the Bible to prove those dogmas.

Mark speaks of the same subject that Matthew does in chapter 10 and verses 1 - 12, when he quotes Jesus as saying, in verse 8 *"But from the beginning of the creation God made them male and female."* Jesus was discussing the subject of marriage and divorce when he said these Words, but we will concentrate on this verse only in this essay.

Some pseudo Bible scholars teach that the first two chapters of Genesis are contradictory; that they are only myths to be interpreted loosely more or less as one chooses. It is important to note that our Lord Jesus Christ did not consider them so. To Jesus, the first two chapters were divinely inspired Scripture with no shadow of contradiction. Note that Jesus used the first chapter and the second chapter in what He said here. He said, *"But from the beginning of the creation God made them male and female."* That is Mark 10:6, and is based uponGenesis 1:27 which says,

"So God created man in his own image, in the image of God created he him; male and female created he them."

In Mark 10:7 and 8, Jesus said,*"For this cause shall a man leave his father and mother, and cleave to his wife; And they twain shall be one flesh: so then they are no more twain, but one flesh."*

In saying that, the Lord was referring to Genesis 2:23,24 which says,

"And Adam said, This is now bone of my bones, and flesh of my flesh: she shall be called Woman, because she was taken out of Man. Therefore shall a man leave his father and his mother, and shall cleave unto his wife: and they shall be one flesh."

The first two chapters of Genesis are not contradictory, and neither are they allegories or myths as some men claim. Let God be true, and every man a liar.

"For ever, O LORD, thy Word is settled in Heaven. Thy Faithfulness is unto all generations: thou hast established the earth, and it abideth."

God created humans as male and female. In fact, all of living creation is male and female except for a very few plants and animals that cannot be positively identified as either male or female. All vertebrates are certainly male and female, and God made creatures like this for a reason. The reason was for the reproduction of the different kinds. Not only were animals created male and female, but they were created in such a way that they could not, or would not, interbreed. Creatures breed with their own kind, and produce their own kind. This is a biological fact that cannot be disputed by observation or test. No one has ever witnessed anything else. This truth is part of the order of nature. That means, of course, that evolution is not biologically possible. Creatures do produce offspring with genetic differences that we call mutations, but mutation cannot possibly account for the trillions and trillions of changes that would be necessary for one kind of creature to become another kind of creature.

Humans were allowed to enjoy the reproductive act for pleasure. Animals are driven by body chemistry and seasons to engage in the reproductive act, but humans were given the privilege of enjoying the behavior as a part of the marriage relationship. The bodies of all creatures were designed so that reproduction could take place only at certain periods, and the female's body changes regulate the time when conception could occur. Animals seem to be completely oblivious to any reproductive drive until the female's body produces certain chemicals.

Humans are preoccupied with reproductive behavior. Sexual immaturity is a root cause of the severe problem of divorce. Divorce is a leading cause of **cultural** failure because it is so damaging to children. Children should be more carefully protected by society by making divorce virtually impossible.

God made them male and female in many more ways than just the reproduction machinery. Male and female chemistry is different. Male andfemale emotions are different. A female is geared to the bearing and care of babies, while the male is geared to the protection and providing for the young. I have made the statement more than once that men are born to die. There is no such thing as unisex, and never will be. A happy person is one who knows what gender he or she is, and understands what roles they are to play, (they stay in their place) and are content to play that role. A man may feel happy playing a woman's role, either in occupation or in some other way, but he will not have the satisfaction in life that he would have had if he had played a role more in keeping with what he was created to be.

"That is just your unwelcome opinion," some will say. Perhaps. But our text says that God made them male and female, and it should follow that a person will be happier behaving according to what God made him. A woman who holds the steering wheel of a big truck rather than a little baby is going to be less satisfied with life than one who raises a child. Yes, I know, a woman can drive a truck and have a child, too, but ask the child where he would rather have his mother be. An unhappy child cannot make its mother completely happy. God made them male and female.

The state of Hawaii is on the verge of changing American culture in a radical way. It was announced last night that a court in that state just struck down the state's laws against sodomite marriages. The subject is almost too revolting to discuss, but we have it thrust upon us. **Paul** said in I Corinthians 5:1 that there were some things that were not even named among the Gentiles, but it seems that we have come to a day of such vile concupiscence that we are forced to talk about everything. The court said that the state had not presented evidence to show why sodomites should not be permitted to marry. The court,of course, did not ask anyone to provide evidence to show why they should be allowed to marry.

Saints do not hate sodomites. Saints hate the practice of sodomy, and the results of it, but the sodomite is a soul that God loves, and died for. I don't profess to love sodomites - I don't even personally know one, but I strongly protest that I do not hate them. The soul of the sodomite is precious to God and therefore it is to me. My message is a warning against the practice of any sin, and an invitation to the sinner to receive the Gift of God, which is eternal life through Jesus Christ our Savior. Christ has already died for the sinner.

TREADING ON SERPENTS

We are studying the references in the New Testament that refer to passages found in the first three chapters of Genesis. The purpose of doing this is to show that the first three chapters of Genesis are divinely inspired, plenary Scriptures with no contradiction among the first two chapters, as some wrongly assert. It may well be that Abel was the writer of the first four or five chapters of Genesis (or Seth), but whoever was the writer, God was the Author, and God is not the Author of confusion, nor is He a deceiver, and so these chapters in Genesis are God's Word, and they are the essential foundation for the rest of the Bible. No one in his right mind would walk through a pit of poisonous snakes. No one in his right mind would go off to a foreign land to live among people who do not care about him. Yet, people could joyfully and safely do both under God's care.

Evolutionists take great pleasure in calling Creationists, and all Saints,for that matter, "ignorant." This is one of the chief weapons they turn upon us. Most **evolutionist**s are highly educated in science, with positions in universities, and therefore have enormous prestige, and power over the minds of people. Having great knowledge, and a long history of education does not make a person wise (Job 32:9). Common sense is essential to being well balanced, and capable of rational thought. Many educated people have neither common sense, wisdom, nor good judgment. This large group of people are those God calls fools in Psalm 14:1 and 53:1. In Psalm 14: 1, God says,

"The fool hath said in his heart, There is no God. They are corrupt, they have done abominable works, There is none that doeth well"

The work of the evolutionist is corrupt, abominable, and deceptive, according to the assessment of the God of creation Who knoweth the heart and thoughts and motives of every man. My own heart is incapable of hatred because it is a redeemed heart. I have no personal malice against any man's soul. Neither does any Saint who understands what he is, according to the Scriptures. However, when the **evolutionist** makes degrading remarks about the creationist, he is voicing his own personal opinion formed in his own unredeemed heart, and it comes forth with malice. Evolutionists have called creationists names like "yahoos," and denied the creationist his credentials without reason.

The creationist cares for the soul of every evolutionist, which means the creationist would have the evolutionist receive all the benefits of being a child of God. Saints are able to love their enemies because Jesus commanded us to, and His commands are not impossible or grievous. Jesus said,

"...Love your enemies, bless them that curse you, do good to them that hate you, and pray for them which despitefully use you, and persecute you." (Matthew 5:44).

Love is a spiritual matter, rising above earthly considerations, and by it a human can love his enemies. I John 1:7:

"Beloved, let us love one another: for love is of God, and every one that loveth is born of God, and knoweth God."

The evolutionist has not only a scientific problem, he has a spiritual problem, and his scientific problem is nothing compared to his spiritual problem. An evolutionist is an evolutionist because he has a wrong relationship with his Creator. He has rejected the existence, or at least the authority, of his Creator in spite of all the good reasons his Creator has given him to accept Him. Evolutionists scorn and mock the Bible, and ridicule the idea of a Heaven and living forever. All of these assertions can be documented from the writings and speeches of evolution philosophers,

If there is someone listening who would like to challenge what I say, then I would be glad to meet you anywhere, any time, under any conditions for apublic debate on the evolution/creation question, and discuss anything you wish in any forum you wish, and I will do it at my own expense. If you feel that a country preacher is not a worthy opponent for a debate, I will gladly provide
others who have indisputable credentials as scientists who will gladly meet you on the same conditions I have listed above.

The first 24 verses of Luke are very interesting and intriguing. So many wonderful things, priceless jewels, lie just underneath many of the Words that are recorded in the Word of God. Blessed is he who takes the shovel of hard study in prayer and digs into the Word of God What a gold mine it is! The Bible is abook for the scholar as well as for the little child, and all who come into its sacred pages with spiritual hunger, and expectation, and who come with seeking heart, will find beautiful things worthy of the heart of the seeker.

In Luke 10, the Savior sends out 70 unnamed disciples who are knowntoday only to God. He gives them instructions which the evangelist today would do well to memorize and follow. Great wisdom is revealed in the Lord's instructions to these seventy evangelists. What an evangelistic campaign ensued! The 70 returned with hearts bursting with joy for what they had been able to accomplish by the power the Lord Christ had given them. So great was their joy, that Jesus Himself was affected by it, and verse 21 tells us that "...*Jesus rejoiced in spirit...*".

My own heart rejoices to think that perhaps in some small deed I have done that I have given Jesus cause to rejoice. I know I have given Him amplecause to weep. How good is the individual who considers his actions beforehand to know whether or not they will cause the Savior sorrow or pain.

"And griev not the holy Spirit of God, whereby ye are sealed unto the day of redemption."

How careless we are about the feelings of Jesus.

After their return in victory, it seems that the Lord gave them additional power, for in verse 19, He tells them,

"Behold, I give unto you power to tread on serpents and scorpions, and over all the power of the enemy: and nothing shall by any means hurt you."

What power! and what a charge! Jesus gave these seventy evangelists power over "*serpents and scorpions*." The Word "serpent" is used by the Lord because these seventy understood that the Master was referring to Satan himself, for it was the scheming serpent who approached Eve in the garden of Eden, and deceived her. The Word "serpent" has been applied to Satan in the figurative sense for 6,000 years by all peoples, because it was the serpent, one of the most beautiful, if not the most beautiful of all creatures, that Satan used in his deception. The serpent must have been a willing subject of Satan for that purpose for God literally **curse**d the serpent with an everlasting curse, never to be lifted.

The serpent may well have been created to walk upright like human beings, because his curse, that of being reduced to crawling on his belly, wouldhave made him the opposite of what God created him to be. Men, and especially women, of course, have always had an aversion to snakes - it's only natural. But we should not take it upon ourselves to eliminate snakes. God has made it clear that vengeance is His, and the world does enough vengeance without people of understanding joining in. We should kill nothing that is not a present threat to us or some person. Snakes are very beneficial to us. I would much rather have a harmless snake around the house than a hoard of mice or grasshoppers. Leave creatures alone unless they are a threat or unless you intend to eat the thing.

It is not the serpent that is our mortal enemy, but the one it symbolizes. It is Satan that we need to be careful about. I Peter 5:8 warns us:

"Be sober, be vigilant; because your adversary the devil, as a roaring lion, walketh about, seeking whom he may devour:".

We will probably never meet Satan face toface, for he is not omnipresent like the Lord is, and he can be in only one place at a time. He probably places himself in some high religious or political center, like Baghdad or Rome or Salt Lake City, for we are told in Revelation 2:13 that Satan's seat at that time was in Pergamos, which was a great center of pagan worship, and so we can expect his seat to be in such a place today.

We will have plenty of encounters with the "scorpions," though, because they are the power of air that swarms around us to do us harm and damage as they can. The fallen **angels** are plentiful.

THE FATHER OF LIES

*Y*e are of your father the devil, and the lusts of your father ye will do. ..."

The New Testament has many passages that refer to thefirst three chapters of Genesis. A knowledge of these three chapters is absolutely essential to a full understanding of the rest of the Bible. These three chapters must be understood for what they are - the foundation of the whole Bible, and they are to be understood in a literal sense if they are to bind the Bible together in a unity.

In Genesis chapter 3 we are introduced to Satan. Satan is not named in the chapter, but we know that he is there, speaking through the serpent which must have been a cooperating creature. In Genesis 1:31, we are told that whenGod finished His creation, and looked at it, He saw that it was "...*very good*...". At that point sin did not exist. How could the creation have been very good if sin was in it? Sin brings death and suffering, and there was none of that in God's creation in Genesis 1:31 when He declared it very good. That means that either Lucifer had not been created at that moment, or sin had not been found in him at that moment.

Evidently Lucifer was created sometime in eternity before the world was founded and so before Genesis 1:31, because he was created as a musical angel to orchestrate the music of the spheres when they were created. Ezekiel tells us that Lucifer's pipes and tabrets were prepared in him (Ezekiel 28:13). He was perfect in his ways. There was music during the creation week because God told **Job** that the stars sang when the foundation of the earth was laid. (Job 38:6,7) Lucifer then was in existence when God declared the universe perfect, and that leaves us to believe that sin had not beenfound in him at that time.

Sometime between Genesis 1:31 and Genesis 3:1 must be the time whensin was found in Lucifer. Sin was found in Lucifer when his pride led him to rebel against God. When God cast him out of Heaven, he immediately attacked Eve. Satan knew that God had told Adam that if he ate of the tree of the knowledge of good and evil, that he would die. Satan, then, in enticing Eve to eat of the fruit, knew that he was in a sense, killing her. Satan was a murderer from the beginning. That is what we find in John 8:44 when Jesus said to the Pharisees, *"Ye are of your father the devil, and the lusts of your father ye will do. He was a murderer from the beginning, and abode not in the truth, because there is no truth in him. When he speaketh a lie, he speaketh of his own: for he is a liar, and the father of it."*

Jesus said that Satan is a murderer and a liar. People ought to be careful about calling other people ugly names, even if the ugly name is deserved. God is the Judge of all men, and only He is in the position to name people according to
what they are.

When Satan rebelled against God, he became a sinner, and sinners cannot have a right relationship with God, and those who do not have a rightrelationship with God cannot have a right relationship with man. It is so good to see people being kind to their fellow man. During Christmas people give a great deal of money to the poor. That is good. It is kind. It is commendable. People ought to be concerned about the bodily health and comfort of others.

But a completely right relationship with other people is a relationship which is concerned with the souls of people. Giving money to a poor person is wonderful, but it does not necessarily show a right relationship with people. Many people give money because it makes them feel good, or because they can mark it off their income tax. That motivation does not show a right relationship with man. It is not the need of the poor that motivates most people to give, it is a selfish motive of some sort that drives them to do good things for others.

When Satan became a sinner, he became the enemy of humankind. Satan abode not in the truth. What a solemn warning this should be to us. We have the truth at our disposal. The Word of God is the Word of Truth. The motivation for deserting the truth can only come from the father of lies, who refused to abide in the truth himself. Truth is perfect. Truth is clean and beautiful. Truth is absolute. Truth is an essential element in human interaction. It is an essential element in interacting with the Lord of Truth. One of the sinsthat broke my heart when I was a lad, and brought me to repentance, was lying. It took years for the Lord to heal me of that awful malignancy, but today spiritual truth is more important to me than breath. Satan did not become Satan, though, because of his action toward man. No person will ever go to Hell because of his actions toward man. No person will ever go to Hell because of some worldly transgression. By the same token, no person can go to Heaven because of something he did to or for some other person or persons or institution. No amount of liquor, dope or adultery or theft or murder can send a person to Hell. As awful as those sins are, and they are awful, Hell is too much punishment for such crimes. Hitler didn't go to Hell, if he did, because he murdered the Jews. Karl Marx didn't go to Hell, that is, if he did go to Hell, because he founded Communism.

Evolution is kin to Communism in that it postulates that there is no God. Having begun with nothing, then, evolution has no place to go. Intellectual pursuits are locked up in the gray matter of its believers, and cannot escape personal experience. Reason is fettered to a vacuum and can only go in circles.

People do not go to Hell because of fleshly behavior. People go to Hell because of spiritual behavior. A sinner is lost, not because he sins, but he sins because he is lost. It is not a person's fleshly behavior that is the root of the problem. It is a person's spiritual behavior that is the problem. The reason it is possible for a murderer to go to Heaven from the execution chamber is because aperson goes to Heaven on the basis of his spiritual condition. The reason it is possible for a rich banker to go to Hell from his mahogany office with its walls covered with plaques telling how good he is, is because a sinner goes to Hell because of spiritual reasons, not fleshly, worldly reasons.

A good man can be lost, and a bad man can be saved. Read the story of the Pharisee and the publican in Luke 18. A genuine mark of a saved person is that they do not mention their good works when they testify of their salvation. Of course, another genuine mark of a saved person is that they hate sin, and will not practice it. Satan did not become a sinner when he tempted Eve. He was already a sinner because he had already rebelled against God.

Salvation is being saved from the original sin. The original sin was a sin of rebellion against God. Eve did not interact with any person, or commit any worldly act that caused her to become lost, or separated, from God. Eating the fruit was a worldly act, but eating the fruit was only the symptom of the spiritual rebellion which was the real problem. Eating the fruit was the result of a decision. The decision was made in an instant, and her decision is what separated her from God. Eve fell, or became lost, because she destroyed her right relation with God. That right relationship was her belief in God, and when she doubted God, the right relationship was destroyed. Doubt will destroy any relationship.

Adam likewise became a sinner when he deliberately chose to disobey God. When he disobeyed God, that demonstrated his lack of Faith in God. Man desperately needs to be saved from sin. Liquor, dope, adultery, perversion, murder - all of those things are harmful to man as are a million other acts no nearly so visible. Adam and Eve were the only saved people in the history of the world who became lost.

God does not look lightly upon any lie. *"Thou shalt not bear false witness against thy neighbour."* (Exodus 20:16). Lying is bearing false witness, and if we receive that statement like God meant it, we stand guilty before God, because breaking one commandment is the same as breaking the whole Law. Salvation is being saved from the original sin. The original sin was a sin of rebellion against God. Eve did not interact with any person, or commit any worldly act that caused her to become lost, or separated, from God. Eating the fruit was a worldly act, but eating the fruit was only the symptom of the spiritual rebellion which was the real problem. Eating the fruit was the result of a decision. The decision was made in an instant, and her decision is what separated her from God. Eve fell, or became lost, because she destroyed her right relation with God. That right relationship was her belief in God.

THE MAKER OF ALL THINGS

I know I worship the right God, that is to say, the only God, because I worship the God who made all things. God judged the gods of Egypt, according to Exodus 12:12, and Paul speaks of gods that are nothing, and so there are multitudes of gods, and though they are nothing, yet many men worship them. I have a comfort and a peace in my soul about knowing that I worship the true God - the only God. I am glad that I have this evidence, not only in my heart, but in my mind, for I know that the true God must be the Creator. The God Who made all things, let Him be God and all other gods be nothing.

It must be pointed out, however, that this knowledge is not saving knowledge. Saving knowledge is the knowledge that the God of Creation became a man, and was made to be sin for us, and died for us in our place. Saving knowledge is not salvation, for there could be those who know all of that, and still have not acted in Faith upon that knowledge, and received Christ Jesus as their own personal Savior in repentance and Faith. The knowledge of salvation does not save, but it is essential to salvation for a person must know what to do to be saved.

Acts chapter 14, Paul and Barnabas are in a small town named Lystra. This is the place where Paul was stoned by an enraged mob of religious bigots, and left in a ditch for dead, and this is probably the time when he was caught up into the third Heaven as he told us in II Corinthians 12:2. Paul and Barnabas had fled for their lives from Iconium to Lystra after the unbelieving Jews there stirred up the city and threatened to stone them to death. We are told about this in Acts 14:4, 5, 6:

"But the multitude of the city was divided: and part held with the Jews, and part with the apostles. And when there was an assault made both of the Gentiles, and also of the Jews with the

rulers, to use them despitefully, and to stone them, They were ware of it, and fled unto Lystra and Derbe, cities of Lycaonia, and unto the region that lieth round about:".

As soon as the two men arrived in Lystra, they began again to preach the Gospel, even though it was at the risk of their lives. Beginning in verse 8 we are told that they came across a man who was impotent in his feet, crippled from his birth. Paul gazed at this man and called out, *"Stand upright on thy feet."*, and the man stood up, and began leaping about with great joy. Of course, the people of the city were amazed, and the news spread through the town like the Chicago fire. People began running to the place where the miracle was performed, and the chatter must have been deafening. They were crying out, *"...The gods are   come down to us in the likeness of men."* (verse 11). They called Barnabas Jupiter, and Paul they called Mercurius *"... because he was the chief speaker."* (verse 12).

Personality-worshippers are still with us. Politicians, athletes, clerics, the rich and famous all have their altars and worshippers. It was a tragedy when a young mother named Diana was killed, but the universal worship of her was wrong. It is right to mourn the death of a human being, but wrong to worship any human. To the horror of Paul and Barnabas, they saw the priest of Jupiter leading an ox, and he had the thing all decorated with garlands and greenery ready to sacrifice it to them. Paul and Barnabas

"...rent their clothes, and ran in amongthe people, crying out, And saying, Sirs, why do ye these things? We also aremen of like passions with you, and preach unto you that ye should turn fromthese vanities unto the living God, which made Heaven, and earth, and the sea, and all things that are therein."

When Paul and Barnabas were confronted with this crowd of idol worshipers, they began preaching the Creator. They began preaching the first three chapters of Genesis, because these people must know that there is only one God. The creation message is greatly needed in the world today.

Paul and Barnabas were somewhat reckless in running into this wild mob, set on fire with religious zeal. We remember scenes from Iran when the body of Khoumini was being carried to his funeral. The wild mob dropped his body out of the casket. Religious zeal can be dangerous. Only a few hours after this scene, Paul is stoned and left for dead outside the city. But they are concerned for the souls of men, and the preservation of their own skin is a matterof little interest to them. Why do people put their lives, and even the lives of their families, in jeopardy for other people that ordinarily would mean nothing to them? Missionaries do this all the time. They count their lives but naught for the souls of people they have never met. God's people are the bravest people on the earth. As soon as the two men arrived in Lystra, they began again to preach the living God ought to make us absolutely fearless. What would it matter in eternity even if we are killed in God's service, except that we would win the martyr's crown? Death is only the window through which we fly into the presence of God. *"O death, where is thy sting? O grave, where is thy victory?"* (I Corinthians 15:55).

Paul was alone in Athens a while later, waiting for his friends, and while he waited, he preached. He disputed with the Jews in the synagogue, and in the market with the Gentiles. All men were precious to Paul. Many were persuaded of the truth of the Gospel, and were saved. While Paul was preaching and teaching in the market, some philosophers heard him, and they were arrested by what he was saying. These Epicureans and Stoicks said,

"What will this babbler say? other some, He seemeth to be a setter forth of strange gods: because he preached unto them Jesus and the resurrection." Acts 17:18)

What an experience to preach to people who have never heard the Gospel before! No wonder Paul gloried in preaching the Gospel where it had never been preached. That is why I went to Siberia. I was looking for a place where the Gospel had never been preached. Virtually all of western Russia has heard the Gospel now. I was in Belarus four years ago, and a Russian pastor told me that three different preachers from Alabama had been in his church in the past three months. He had been insulted by them, and he wouldn't even let me in his pulpit. I came home determined to go where the Gospel had never been preached. I went to Siberia.

When I went to the city of Tomsk in Siberia, I found that few people had heard the Gospel, and I don't believe there had ever been an American fundamental missionary there.

The reason it is so good to preach the Gospel where it has never been preached is because people hear strange things when they hear the Gospel preached the first time. It is strange to hear that you can be saved without the approval of some priest. It is strange to hear that you can be saved by Faith without works. It is strange to hear that God loves you, and your soul is precious to God when nobody else ever loved it. It is strange to hear that God created the universe out of nothing.

These philosophers took Paul up on Mars' Hill where he preached to them the Creator (Acts 17:22). When we are confronted by pagan idol-worshipers, we must introduce them to the God ofCreation. It is the Creator who provided the Gospel. Paul said, *"God that made the world and all things therein..."* (verse 24). Let's be sure we are dealing

A sinner cannot understand the Word of God (I Corinthians 2:14). Regardless of the educational level of the individual sinner, he is ignorant in the Bible. Dr. Isaac Asimov was an intellectual, the author of many books, yet he could not understand the spiritual nature of the Scriptures. He wrote two Bible commentaries, but he could write only about the history of Israel and the nations they interacted with. The spiritual undertones of the Word were invisible to him.

The Bible is too complex for an ordinary person to find its deep and wonderful truths. Asimov, in all his fine education, could not understand that "... in the beginning God created the heavens and the earth." (Genesis 1:1). Dr. Asimov was a scholar par excellence, but he was not whole because he was a fool when confronted with the Word of God. The Syrophenician woman was willing to accept God's assessment of her. If He had said she was a Gentile dog, then she must have been. In accepting His assessment of her, she acknowledged Him as a righteous judge. When God says that we are sinners, we must be willing to accept that as the truth. She was no Gentile dog when she held her healthy little daughter in her arms. She was a saint.

The Scriptures tell us in verse 28 that the girl was made whole. Without seeing this child, Jesus cast out the devil that was tormenting her. And she was made whole because that devil had occupied her body, and surely had destroyed part of it.

INVISIBLE THINGS

There are at least four references in the Epistle to the Romans to the first 3 chapters of Genesis. These are 1:20; 5:12; 8:21,22; and 16:20. There are probably some other passages that others would consider in this list. For our study today we will select Romans 1:20. Let's read the verse together, and include verse 19 to improve our understanding of the context:

"Because that which may be known of God is manifest in them: for God hath shewed it unto them. For the invisible things of him from the creation of the world are clearly seen, being understood by the things that are made, even his eternal power and Godhead; so that they are without excuse:".

It is interesting to note that the second Law of **thermodynamics** applies not only to matter and all of nature, but it applies to social, religious, and philosophical matters as well. As we have discussed several times on this broadcast, the second Law of thermodynamics is the scientific term that identifies the forces of nature that causes everything to wear out, cool off, and die. Everything becomes less complex over time. Heretics who study the Bible to find fault with it, or to reduce it to a purely human product, claim that the Jewish and Saint religions, as they call them, evolved upward from ancient pagan religions. They cite ancient Babylonian idol worship, for example, as being one of the religions from which Judaism evolved. Those who can gain eminence in a particular field are not above taking advantage of their prestige to propagate their own philosophy.

This is pure nonsense, just like all the other preposterous claims evolution philosophers make. Saintity, and the salubrious, wholesome ethics it involves could never have developed out of pagan religions. Man does not have the capacity to learn about God, and the ethics He provided, out of the natural resources of his simple brain. The truth is that pagan religions are a corruption of man's mind, deranged by sin. One of the most revolutionary ideas ever placed before the human mind was Jesus' command to

"...Love your enemies, bless them that curse you, do good to them that curse you, do good to them that hate you, and pray for them which despitefully use you, and persecute you;" (Matthew 5:44); and

"...Thou shalt love thy neighbour as thyself." (Matthew 19:19). Man would never have thought of such a thing. If we are to hold our neighbor in such esteem, surely we are to hold our brother in greater esteem, especially those of the household of God.

"As we have therefore opportunity, let us do good unto all men, especially unto them who are of the household of Faith."

The Bible is right and the **evolutionist** is wrong again. Our passage clearly states that man knew God at the beginning, even understanding Him to be a triune personality. The next verse in this great chapter says,

"Because that, when they knew God, they glorified him not as God, neither were thankful; but became vain in their imaginations, and their foolish heart was darkened."

God created the universe at the beginning, and man knew it. Doubtless, Adam wrote books about it, but man chose not to retain God in his knowledge. This is the story of human history. It is the story of individual history.

"... the invisible things of him from the creation of the world are clearly seen ..."

was a very difficult passage until I came across some instruction from a scientist who is a Bible scholar. The man of whom I refer is Dr. Henry Morris, whom I consider the most influential man in twentieth century. Dr. Morris points out the scientific tri-unities of matter, time, and space[1] and shows how thesescientific truths are consistent with the biblical teaching of the Word of Godconcerning the person of Jehovah. He is the triune God: Father, Son, and Holy Spirit. I get distressed sometimes by the lack of interest of people in this important - vital - subject. Even among churches that are thought of as being fundamental there are many whose pastors do not hold to the literal meaning of the creation account. There are many others who feel that the creation story is more or less irrelevant. And others who feel that it is just a divisive issue betterleft alone. The universal teaching of evolution religion must be countered, and it can only be countered by a literal account of creation.

Anyway, let it be known that there is no way to understand Romans 1:19 and 20 unless you understand them from a scientific standpoint. The entire creation work of the Lord was absolutely scientific. It was miraculous in that the Laws of physics did not exist until God created matter, and then He had to miraculously create those Laws to govern the behavior of matter. At the beginning, man was much more intelligent than he is now, and his understanding of the movement of the Heavenly bodies, and many other concepts, was greater than it is now.

A great deal of interest is being shown in the study of the signs of the zodiac. Some imminent scholars believe that these signs, probably referred to in Genesis 1:14, revealed the redemption story. Early men had the intelligence to read the signs. The quick discovery of iron, and the methods of making alloys point to man's great intelligence.

During my conferences I give a Sunday school lesson on Sunday morning with the boys and girls in the front of the church, and the adults looking on. Using a small board, I have boys and girls make certain marks on the board.With this method, I teach how that

"The Heavens declare the glory of God; and the firmament sheweth his handywork." and the audience is amazed by what they learn. For example, I point out that we live in a three dimensional universe. Three dimensional meaning about the same as "triune."

God created the three dimensional universe ("three dimensional" means length, width, and height) to point man to Himself - the triune God. God created matter in three states: liquid, solid, and gas - pointing man to Himself, the triune God. God created time in three elements: past, present, and future - pointing man to Himself, the triune God. On the third day, God created plants in three classes: grass, herbs, and trees. God created animal life in three classes: fish, land animals, and birds. Altogether in Genesis, God created only three things: matter, physical life, and spiritual life.

There is a debate in progress now on the Nova Internet online[2] between Professor Phillip Johnson, of the University of

California at Berkley, and Professor Kenneth Miller of Brown University. Dr. Johnson is a law professor and author of *Darwin on Trial*, and Dr. Miller a biologist, and author of a biology textbook. In one of his letters, Dr. Johnson writes, "The heart of the Darwinian religion is the claim, advanced in all the textbooks, that evolution is an undirected and purposeless process that produces humans by accident."

In his assessment of Darwinian evolution, Dr. Johnson gives us the reason for the present low state of American morals. If humans are produced for no purpose, then there is no purpose in life, and life becomes virtually worthless. Man made a grievous mistake in not retaining God in his knowledge. Lack of light makes things invisible. That's the reason people love to riot in the night. Under the cover of **darkness** all sorts of bad things can be done, and people think they are getting away with their crimes. They mock at God but God will not be mocked. We are told that the night and the day are the same to Him.

This is Paul's testimony before Agrippa. He is repeating Christ's commission to him. The Gospel Paul will preach will deliver the Gentiles out of darkness into light where they will be able to see the Light which will deliver them out of darkness where Satan had them bound in sin. What a glorious thing to be able to see. My granddaddy Kennedy was blind physically. I watched him grope. Every beautiful thing around him was invisible to him. He could not see the beauty of a butterfly on the dahlias before his front porch. He could not see my grandmother's lovely face when she was young. He never saw his three youngest children. He could not work. He cut the grass by taking a handful, and carefully cutting it with a sickle. The Gentiles lived in darkness, bound by Satan.

---

1. Henry M. Morris, *The Biblical Basis for Modern Science* (Grand Rapids, Michigan: Baker Book House, 1984), pp. 59-66.
2. http://www.pbs.org/wgbh/pages/nova/odyssey/debate

MAN-MADE FOSSILS

There are at least ten references to subjects in the first three chapters of Genesis, found in the first epistle to the Corinthians. A great number of pages in books have been written about these verses, and I am sure that many more could be written. Quite a large number of pages have been written about I Corinthians 15:21, but we will proceed to say a few more things about this important verse of scripture. Let's read the verse.

"For since by man came death, by man came also the resurrection of the dead."

This statement is so clear that it serves as an anchor for other passages which may be more difficult. **Fossils** could not have been formed before man sinned.

It is sad that so many people have a bad opinion of God. Many people today insist that God is responsible for all the suffering and death in the world, and demand that God stop all the pain and suffering in the world. They insist that there was pain and death in the world before there was a man on Earth, and this is accusing God of bringing sin and death. Many **evolutionist**s claim to believe in God, but their insistence upon believing that **fossils** were formed millions of years ago is evidence that they do not believe in the true God. Here is why. Evolutionists say that the oldest life forms on Earth are fossilized in the crust of the earth in rock that they say is 500.000,000 years old. This strata is called the Cambrian period, and in Cambrian rock we find fossilized trilobites,and other very small creatures. Evolutionists say that man did not appear on Earth until about a million years ago. That means that for 499,000,000 years creatures died before man appeared on the earth.

Each person must decide between evolution and the Bible. According to the evolutionists' time scale, man could not have been responsible for death. That is a direct contradiction of the Word of God. God says that death came by man. Evolution is a pernicious idea that has done nothing but cause misery and heartache to the human race. Evolution has never produced one single thing of value to mankind. Nothing. It is a philosophy that causes people to think of people as animals, and causes a great amount of pain and suffering and death. It is not science. Science is beneficial to humankind, and actually has a biblical mandate. A man on the Internet wrote that he believed there is more evidence for the existence of the devil than there is for the existence of God. You have to admit that there is some validity in that statement. There are more sinners than saints in the world. The wicked have more power and wealth than the righteous. Death and decay are all around. Man loves **darkness** rather than light. A person gets an idea like the above from university classrooms where evolution is the daily dose of dogma. Teaching evolution to the exclusion of others is indoctrination.

Mr. **David** Buckna is an Internet friend who lives in Canada. He is a school teacher, and author, and a defender of the creation account in Genesis. He wrote, "Remember back in November '89 when the Berlin Wall came down? After three decades, a section of the Wall was torn down by German citizens in a matter of hours. By 1992 most of it was broken up for use in roadbeds and other construction projects. The same sort of thing is beginning to happen with evolution, and it's starting to crumble a lot faster than many might think."

Even though I am afraid Brother Buckna might be a little too optimistic in his appraisal, I sincerely hope, for mankind's sake, that he is right, but people don't give up their religions very easily, especially when those beliefs are fanatic beliefs. After giving some reasons for his opinion, he goes on to say, "In the not-so-distant future, when someone of the stature of a Stephen Jay Gould or Carl Sagan holds a press conference to announce he has finally reached the conclusion that evolution is scientifically bankrupt, other scientists will quickly follow suit. It'll resemble rats deserting a sinking ship."

Our passage in I Corinthians 15:21 refers back to Genesis 3:19. Check out these verses in your Bible.

God told Adam in the garden that his disobedience would cause him to die. He would return to the dust from which he was taken. Adam's sin caused his own spiritual death immediately, for no longer would he enjoy the fellowship of God as he had before. His sin caused the physical death of every man and animal on Earth.

No creature ever died before that moment. That is abundantly clear from our Corinthians passage, as well as from Romans 8:12, and other passages. No creature died before Adam sinned, and therefore no fossil could have been formed before Adam sinned. This truth from the Bible is abundantly supported by all of the evidence of science. One writer on the Internet stated that he believed in God, but he also insists that the fossil record proves evolution to be true. I would hate to think I believed in a god who would have to use the deaths of trillions of animals to bring life on Earth to its present state. Read II Timothy 1:10.

Jesus the Life-giver, has abolished death for those who come to Him through Faith in His blood.

"Knowing that Christ being raised from the dead dieth no more, death hath no more dominion over him." (Romans 6:9).

It was man who made the **fossils** - not God. It was man's sin that brought death into the world - not a hateful God. Death is the result of man's sin. Those who would accuse God of being a tyrant for not just overlooking man's one little sin are ignorant of the consequences of such an action by the Lord. I am glad God kept His word. I am not glad that so much death and suffering has resulted, but I am glad that God kept His place. What sort of world would it be if Adam and Eve were gods as God is?

There is good news in our verse. Not only does it tell us the bad news that death came by man, but it tells us the good news that by another man *"...came also the resurrection of the dead."* It was not God's choice that death came into the world, but it was God's choice that His only begotten Son came into the world to destroy death. All must die because of Adam's sin, but all can be resurrected because of God's righteousness. Sin will be with us as long as sin reigns, and so evolution will be with us as long as sin reigns.

Here read Romans 14:11

God is *"...the resurrection, and the life: he that believeth in me, though he were dead, yet shall he live."* (John 11:25).

In his excellent book, *Bones of Contention,* Dr. Marvin Lubenow relates the story of a Neandertal skeleton found in southern France in 1908. The skeleton was the most complete skeleton that had been found up to that date. The remains were turned over to Marcellin Boule of the National Museum of Natural History in Paris to be reconstructed. Boule was a famous paleontologist, but in this instance he came to his task with some serious preconceptions. He believed that the Neandertals could not have been in the line of evolution to modern humans, certainly not Frenchmen. He was certain that Piltdown Man, and the Cro Magnon people were the ancestors of modern man. Piltdown Man was Boule's choice for the honor of being in the line of modern man. But Pildown Man, discovered in England by Charles Dawson, was later exposed to be one of the most famous of all the hoaxes of science.

Boule proceeded to construct the Neandertal skeleton to suit himself, and to prove his theory that Piltdown Man was the ancestor of man, not the Neandertal. Boule published his report on the skeleton between 1911 and 1913, and it stood for forty-four years until 1955 when two doctors named William L. Straus and A. J. E. Cave, looked at the skeleton. They immediately saw the problems with the reconstructed skeleton, and published a paper exposing it in 1957. Neandertal did not look and walk like an ape at all. He was straight and tall and intelligent. Boule lied for years until he was caught. His man-made fossil was exposed for what it was – a lie.

The Field Museum of the Natural History in Chicago had its own display of Boule's Neandertal skeleton. They were not ignorant of Straus's and Cave's discovery of the Neandertal skeleton hoax, yet they left their display of the skeleton in place for another twenty years! They at last moved the display to the second floor, unchanged, where they placed it beside the skeleton of the huge Brontosaurus skeleton. Here the lie was perpetuated, but doing more harm than before since here it is seen by more children. The museum, one of the world's best museums, placed a sign on its bogus display reading, "An alternate view of Neandertal." Lubenow says, "It was not an alternate view. It was a *wrong* view of Neandertal."[2]

---

1. David Buckna, "Evolution: It's Collapse in View?" (Internet address, dabuckna@awinc.com: 12-10-96).

2. Marvin Lubenow, *Bones of Contention* (Grand Rapids: Baker Books, 1992), pp. 36-41

THE WOMAN'S SEED

We now come to the epistle to the Galatians as we continue our study of the New Testament references to the first three chapters of Genesis. How grand is this study! How marvelous to know that God has placed His divine seal of approval upon the first three chapters of Genesis by weaving them as a golden cord all the way through the Holy Writ! My soul is thrilled to its innermost vaults by such grand and glorious knowledge. The power of the ages flows from the pages of the Bible like a great dynamo, electrifying every believing heart with joy unspeakable and full of glory. Such knowledge is too wonderful for me. I must seek relief from the ecstasy of it by sharing it with others. *Galatians 4:4 says,*

"But when the fulness of the time was come, God sent forth his Son, made of a woman, made under the Law."

What a waiting period there had been before the Savior was born! Forty, maybe fifty centuries passed after God promised the woman's seed in Genesis 3:15. Eve thought that Cain was the promised Redeemer, but alas, he was a murderer. Preparation must be carefully made, and the human race must increase in numbers, and its cup of iniquity must become full before the promised Seed would come into the world. But at last, the time did arrive. It was 2,000+/- years ago. God did send His Son, *"...made of a woman..."*. Some may wonder why God did not say "virgin" there, but we must not allow our wondering to become skepticism or unbelief. A woman can be a virgin, whereas a virgin is not necessarily a woman. God has already told us that the mother of Messiah would be a virgin.

"Therefore the Lord himself shall give you a sign; Behold, a virgin shall conceive, and bear a son,...".

The Word of God makes it abundantly clear that Jesus was virgin-born. It would have to be so. It would go without saying, anyway, that the Son of God would be born of a woman who would be untouched by a man. To believe that Jesus is the Christ is to believe that He was born of a virgin.

Some have questioned the deity of Christ. Both the Old Testament and the New declare Him to be very God, and it is evident that no mortal could have done what Messiah did.

Isaiah told us that Immanuel would be born of a virgin. 7:14 says,

"Therefore the Lord himself shall give you a sign; Behold a virgin shall conceive, and bear a son, and shall call his name Immanuel."

The Lord gave this sign to one of the worst kings Judah ever had, but in actuality, it was to the people. God could not allow Judah to be annihilated, because it was from Judah that Messiah must come. But no one appeared who was worthy or capable of carrying the weight of the name Immanuel - God with us. Ahaz died. The people then living all died. Isaiah died, and still no one was born of a virgin, and no one who could carry such a name as Immanuel had appeared.

Isaiah did not remove the prophecy from his book before he died. He left it there, knowing that it was the Word of God. Other great men came and went. Nehemiah came and lived, but he was not great enough to carry the name Messiah. There was Josiah, Judas Maccabeaus; the great Daniel came on the scene, and Hosea, and Micah, but the name was too great for them. What God has prophesied through His holy prophets can no more be held back than a mighty river or the sun in its course, or the moon in its orbit. What God wills, will come to pass it its time.

Sir Thomas Malory wrote *King Arthur* the story about how Arthur was discovered and declared King of Britain. Arthur, the only son of the serpent dragon, had been stolen by Merlin as soon as he was born, and had grown up unknown. The kingdom was in confusion, and many of the great barons coveted the crown of England. The story goes that the Archbishop of Canterbury invited all the barons and knights to London on Christmas eve for a Christmas day service to be held in the greatest church.

When they came out of the church, they saw in the churchyard a great marble stone with a steel anvil sitting on it, and in the anvil there was stuck a gleaming sWord. Emblazoned on the sWord in letters of gold were the Words, "Whoso pulleth out this sWord of this stone and anvil, the same is rightwise kingborn of all England." The great lords struggled to pull the sWord from the anvil and stone to no avail. Then came the unknown prince, and without so much as a grunt, pulled Excalibur out the anvil. They repeated the trial at Candlemas, at Easter, and at Pentecost, but none had the strength to pull the mighty sWord from the anvil except Arthur, who pulled it out every time with ease. Only the King of England could bear the great sWord Excalibur[1]

At the time of Jesus' birth, there were many who claimed to be Messiah, for all knew when and where the woman's Seed was to be born, but none of them were mighty enough to bear the name "Emmanuel." Indeed, there could be only one Emmanuel - God with us - and He was easily recognized by all who would look. Only Jesus was the virgin's Son. Only Jesus was the Seed of the woman, promised in the garden of Eden so many centuries before. Long had the world waited for the coming of its Creator, and now, garbed in the likeness of sinful flesh, He strode onto the stage of human kind and changed it forevermore. He even divided time into two parts.

Jesus Christ was the comrade of men. Jesus Christ was the healer of men. Jesus Christ was the Teacher of men. But He was much more. He bore the name Emmanuel, and He alone stood among men as God, and King, and Sacrifice. He was God's perfect Lamb, offered for humanity's redemption. Yes, in the fullness of time, God sent His Son, made of a woman, made under the Law, and He was such a Son that God the Father could declare from Heaven to the whole world, *"This is my beloved Son, in whom I am well pleased; hear ye him."* (Matthew 17:5).

Some, in the wickedness of doubt and skepticism, have said that Isaiah did not promise a child born of a virgin, and others have asserted that it doesn't matter anyway.

"...Behold, a virgin shall conceive, and bear a son, and shall call his name Immanuel." (Isaiah 7:14).

Some have proclaimed that the idea that a child could be born of a virgin was no more than a pagan myth. Such wickedness!

- 217 -

Such an insult to the holy Son of God and His Father, and godly Saints who love Him. What an insult to the Father and to the Holy Spirit. If Jesus had a human father as some affirm,then certainly Mary, the only one who could actually know for a certainty, would have confessed it when her Son was being condemned for being a blasphemer claiming to be God. Mother love would have driven her to sacrifice herself for the sake of her Son if He had been conceived by some mortal man, but she stood own soul also, as ancient Simeon had prophesied on that day in the temple, and watched her first born Son die in untold suffering on the rugged cross. Matthew makes it abundantly clear that the Old Testament prophet did indeed mean that a virgin would conceive, for he wrote in chapter 1 and verses 21 and 22,

"Now all this was done, that it might be fulfilled which was spoken of the Lord by the prophet, saying, Behold, a virgin shall be with child, and shall bring forth a son, and they shall call his name Emmanuel, which bein interpreted is, God with us."

The Apostle Paul never mentions the virgin birth of our Lord Christ, because it was a foregone conclusion that Jesus was the Son of God, and it just naturally followed that He would be given his mortal body, and brought into the world by a young woman who had neve been touched by a man, even a husband.

The virgin birth of our Lord is not a subject for debate. It is a matter to die for. Upon this fact we take our stand as fundamentalists, to hold it dear even if holding it requires us to carry it with us to the stake. The birth of our Savior is a personal, intimate matter that only a devout physician like Dr. Luke could be qualified to take pen in hand and write the Words whispered into his ears by the Holy Spirit of God. Only in Luke's training and experience could the Lord find the needed Words. The reading of those Words should be a matter of equal reverence and worship. Oh, come, let us adore Him.

---

1. Edward Hastings, *The Speaker's Bible* (Grand Rapids, Michigan: Baker Book House 1987), vol. 4. p. 67,68.

THE FELLOWSHIP OF MYSTERY

In Ephesians there are two rich references to the first three chapters of Genesis. Ephesians 5:30 and 31 refer to Genesis 2:24, and Ephesians 3:9 refers to Genesis 1:1. It is Ephesians 3:9 that arrests our attention today, and brings us to apply our minds, watched over by our hearts, to the thoughts residing therein. Let the thoughts and intents of our hearts always be in tune with the living God, the Creator, Who provides our next breath. The desire of the heart should be to honor Him. *"I will bless the LORD, who hath given me counsel: ..."* (Psalm 16:7). Ephesians 3:9 says,

"And to make all men see what is the fellowship of the mystery, which from the beginning of the world hath been hid in God, who created all things by Jesus Christ."

God did many things in Christ. How could God do things in Christ Who is also God? I am Joseph Kennedy. Everything I do, Joseph Kennedy does it. I am Joseph Kennedy Body; I am Joseph Kennedy Soul; and I am Joseph Kennedy Spirit. Joseph Kennedy, though old and tired, does many things. Joseph Kennedy began this day today at 3 AM after being up until after 10 o'clock last night. I am three persons. I am Joseph Body; I am Joseph Soul; I am Joseph Spirit. In preparing this essay, Joseph Kennedy in Joseph Body opened books and looked at pages. Joseph Kennedy in Joseph Soul interpreted what Joseph Body saw on the pages of books. Joseph Kennedy in Joseph Spirit prayed that God would use all that Joseph Kennedy is to glorify Himself. Joseph Kennedy can do three things atone time because he is three persons.

II Corinthians 5:19 says, *"To wit, that God was in Christ, reconciling the world unto himself..."*. God reconciled the world unto Himself, and He did it in the person of Jesus Christ. Of course, this is part of the mystery that shrouds the **Godhead**. It is relatively simple to me, but yet it is not simple, and mysimple understanding leaves a great many things to wonder about. God does not reveal Himself to man in all the intimate details of His person, because God is God, and there is a sphere in which He moves that man cannot enter into.

Familiarity breeds contempt, somebody said, and a veil of mystery is apractical garment for all of us. We should never lay bare the inmost secrets ofour heart. There should always be secrets that give others cause to wonder, and to respect us for.

The Bible contains mysteries. This is one of the reasons we know the Bible is the divinely inspired, plenary Word of God.

"Who can find out God, or know the Almighty unto perfection?" "No man knoweth the Father save the Son." "No man hath seen God at any time." "No man can see God and live."

What do you expect when you come before the King? Did you think He would tell you all about his private affairs? The Bible brings before the human mind a great number of mysteries, but if the Bible were removed, it would not eliminate the mysteries. How dull life would be without mysteries!

Our verse speaks of *"...the fellowship of the mystery,"* and that places all of humanity in a communion of wondering, exploring, studying, exchanging ideas to the end that we might lift somewhat the mystery. The existence of the mystery is not altogether a wall which is not to be approached or peeked behind. The Bible does not ask to be accepted because it takes away all mysteries. The Bible has many mysteries, but we are allowed to know enough to know that the mysteries will all be solved bye and bye. A mystery is not the same thing as a secret. It should a concern that nothing can be kept secret from God. Psalm 90:8:

"Thou hast set our iniquities before thee, our secret sins in the light of thy countenance." A wise person will live daily with the knowledge that God is privy to every secret of his heart and soul.

God told the Jews over and over again in the Old Testament that the Gentiles would seek after their Messiah, but never did it occur to the Jews that Jews and Gentiles would become one in Christ, and be called the church. You could probably say that a mystery is a secret which will at some point in time be made manifest. Mysteries are discerned by people who will open their eyes to the evident truth that is before them. **Evolutionist**s will not open their eyes to the truth of the Scripture or to the scientific evidence produced by theirown labors. Scripture or to the scientific evidence produced by their own labors. Creation is no secret, but it is a mystery to those who will not open their eyes to the truth. The Bible is a Book of Light, but where there is light, there are shadows, and these shadows are the mysteries. Shadows are dispelled when light is shined upon them. The light of the Bible can be shone upon its own shadows, just as a lamp can dispel its own shadows by a little human effort. You just move the lamp.

Mysteries make life interesting. There was timeless eternity when God was alone in the universe, but as soon as He created the first atom, there had to be time. There can never be timeless eternity again, for as a man walks down the street of glory, it will take him a certain number of units of time to walk from one place on that street to another. On Earth there is only one dimension of time, but in Heaven there may be many dimensions of time. It may be that God exists in many dimensions of time at the present. This is a great mystery, and though some may not care to dwell upon questions that seem to have no answer, life forsome people is made fuller by wrestling with hard questions.

Mr. Herbert Spencer, whose mind was potent but dark, said, "The Power which the universe manifests to us is utterly inscrutable." Mr. Spencer reveals in this thought just how much darker are the mysteries of life to the unbeliever than they are to the believer. The "Power" which the universe manifests to us is mysterious, but it is by no means "utterly inscrutable." The "Power" which theuniverse manifests is spiritual power, and those who look at the universe as if it were nothing but matter, surely will never comprehend the "Power." "Power" is understood by those who can discern matters in a spiritual context.

God gives power to His children.

"For God hath not given us the spirit of fear; but of power, and of love, and of a sound mind." "And Stephen, full of Faith and power, did great wonders and miracles among the people."

Self-worship likewise fails to shed a beam on the "inscrutable power." Man, who is programmed to believe that man has an explanation for everything, can in nowise penetrate the mysteries of human kind, the Lord, or the world around us. Total blindness can result from thinking that man is supreme, or that matter is supreme, or that time is supreme. Elihu, the young man who joined **Job** and his three tormentors at their dying campfire, said, *"Behold, God is great, and we know him not, neither can the number of his years be searched out."*

Elihu may or may not have spoken by divine revelation. We know not. We do know that **Job**'s three friends did not speak by divine revelation, though their Words are recorded by divine revelation, because God said they did not in Job 42:8. Nevertheless, we know that at least part of what Elihu said is true. Elihu spoke as a materialist when he said that we know not God. Though God is shrouded in mystery, there is a great deal that we can know about Him because He has been pleased to reveal much of Himself to us through His only begotten Son, and His Word.

The mystery of human life will never be understood by those who see the humans as nothing by evolved matter. The mystery of human nature in all its complexities can never be understood, or in the least comprehended, as long as it is considered to be nothing but atoms of matter. What then, can be the purpose of looking into these matters? Man cannot solve the problems besetting the human race without understanding that man is more than atoms in a highly structured form.

A mystery in the New Testament identifies truth that has been hidden, but now is revealed through the Holy Spirit. The mystery is not hidden from God's people primarily, but from Satan. Had Satan known that Jesus Christ would be born (I Corinthians 2:7; Colossians 2:2,9); that there would be a Gospel (Ephesians 6:19); that there would be Christ and His church (Ephesians 5:32); that there would be an indwelling of Christ (Ephesians 1:27); that Jews and Gentile believers would unite in one body (Ephesians 3:1-12); that there would be a kingdom of heaven (Matthew 13); that there would be a Rapture (I Corinthians 15:51) he and his demonic host would have tried to prevent it all from coming to pass. Therefore, God waited until the church was well-established before revealing these mysteries.

RENEWED IN KNOWLEDGE

As we move through the New Testament studying some of the passages that refer to the first three chapters of Genesis, we come to Philippians, but alas, Philippians does not have a substantive reference to any passage in the first three chapters of Genesis. Philippians 4:18 does refer to Genesis 4:4 and 8:8, but our study is confined to verses that refer to the first three chapters. At the beginning of this study I pointed out that Philippians, I Thessalonians, Philemon, Jude, I Peter, James, II John, and III John have no reference to the first three chapters of Genesis.

And so we pass on to Colossians today. The epistle to the Colossians has two references to Genesis 1 - 3 in it: chapter 1 verse 16, and 3:10. Colossians 1:16 is a key passage concerning creation, and it says,

"For by him were all things created, that are in Heaven, and that are in earth, visible and invisible,whether they be thrones, or dominions, or principalities, or powers: all things were created by him, and for him:". Colossians 3:10 says, *"And have put onthe new man, which is renewed in knowledge after the image of him that created him."*

For our study today, let's select this passage for consideration. The new man. Old men can be made new men. Old men can be new men. That's the good news of the Gospel. A man either has the blood of Jesus Christ on his heart, or he has that blood on his hands. One way or another, man must be accountable to his Creator. Those are serious Words "give account." Whether one is standing before his daddy, his boss, or his Maker. Take heed. Blood on the soul by Faith in the Lamb of God makes a new man. The new man has the potential of being full of joy unspeakable and full of glory. The new man has the capacity to be wiser than the most educated man in the world. The new man has put off the old man with his deeds. No more works for salvation, but works for the glory of God. The new man is the man who can be pleasing to God. The new man is a sinner saved by the grace of God.

What are the deeds of this old man? They are listed in verse 8: anger, wrath, malice, blasphemy, filthy communication, lying.

Praise God for the joy of being rid of them. What glory to know that the new man is not in bondage to those festering sores! What a perversion for the new man to be guilty of those things. What an ugly sight to see the new man harboring malice in his heart. What a broken, rusted, rotten thing to see the new man pouring filth out of his mouth. The new man *"...is renewed in knowledge after the image of him that created him."* I love knowledge. I have sought it with bitter suffering and sacrifice nearly all my life. But knowledge without Christ is a rubbish heap, doing no more than poisoning the soul. *"...knowledge after the image of him that created him."* is knowledge worth dying for. Oh, to *"...know him, and the power of his resurrection."*

To be *"...renewed in knowledge..."*. At one time man knew that he was created in the image of God. Man is born knowing there is a Supreme Being. But man did not like to retain God in his knowledge, and so God gave him over to a reprobate mind, as Romans tells us, and *"Professing themselves to be wise, they became fools."* The tragedy of human life. Reading the babble on the Internet is evidence that man is truly deranged spiritually, and this spiritual derangement radically affects his power to think and reason rationally. The greatest evidence of this irrational, spiritually corrupt mind is the preposterous idea that there is no Creator responsible for life on this planet. All scientific and common sense reasons against evolution make it impossible to believe in evolution with a rational mind, yet it is the fashion of the day. How did evolution get out of the philosophy class into the science lab? It seems impossible that such a transition could have occurred, yet it did, right under our noses. The blindness of humans is no more prevalent that in this case, where the Word of professors is preferred above preachers. People believe what they want to believe on the basis of who else believes it. The university culture professes to believe in evolution, and so university personnel hold to the idea in spite of the paucity of reason to believe it. Birds of a feather flock together, somebody said, and some birds imagine themselves certain colors just because they like the color. Denying evolution is like denying the Jewish Faith, or Islam. The person who pulls out is ostracized, cast out, and blacklisted. For a college professor to make it known publicly that he has rejected evolution is virtual professional suicide.

Unfortunately, so many churches have made a reputation of being completely unintellectual, that many educated people are ashamed to stand beside them in what they believe. The Gospel is set in a timeless framework, and cannot be changed or modernized. But the preacher who does not recognize the particular problems of the culture in which he lives, and the issues of life in the time in which he lives, is going to find the world ignoring him. If the preacher does not recognize that Genesis is being viciously attacked by every element of our society, he is going to find his people drifting, and his young people uninterested in anything but the entertainment the church provides. Children are blessed when they learn that there is a time and place for different things, but there is never a time and place for certain things. Wise parents can give their children an enormous boost by leading them to understand the times and places. Being casual all the time produces a crumb-bum.

The preacher who addresses the wickedness of the present world, such as what the child must live with in school every day, and their parents on the job every day, will find an attentive audience three times a week. I am a firm believer that when the pastor stands to preach, every member of his church should be under the sound of his voice. I would not allow my children to go to some club meeting somewhere else in the church during the time the Word of God was being preached from the pulpit. There is no substitute for the preaching of the Word of God. We are raising a future church that will be so immature that the future of the church would be in question if it were not God's church under His protection. Parents allow their children to sit somewhere else in church, and have no idea what they are doing during the service. Other parents allow their children to sit when the pulpit says stand; children put their feet up on the pews in front of them, and disrespect the house of God.

In a little booklet I just received in the mail[1], Dr. James M. Boice writes, "In 1987 University of Chicago professor Allan Bloom published a book in which he exposed the demise of education... .

Historically the goal of education has been the pursuit of the good, the true, and the beautiful. 'But,' wrote Bloom, 'that goal is impossible today, because in a culture dominated by relativism belief in absolutes like the good, the true, and the beautiful no longer exists'." Dr. Boise, under a subtitle mentioning "barbarians," attributes the breakdown of American society generally, to "... the cult of self and self-fulfillment to be achieved at the cost of nearly everything else." That is pure evolutionism in action.

Saints are engaged in a struggle to hold American society's head above the water. If we do not prevent the total collapse of our society, then the e vangelistic work of the churches, in fact, the very existence of the Church in America, will be threatened. Saints should never try to influence society in the name of the Church. They must do that sort of work as citizens, for the business of the Church is to win the lost, baptize them, and teach them to win others. It is self-evident that it is essential that the Church be renewed in knowledge. Our ancestors knew how to raise children. We can relearn how to raise them. Many people know how, and are doing it. Others need to examine their own approach to the work, and make some changes.

Education in America has become more secular than it has been in the history of the nation. The knowledge boys and girls received in schools early in the nation's history was geared to the Scriptures, and boys and girls learned honesty and the work ethic from a divine authority. A divine authority is needed, because man's authority is only his opinion, and everybody is entitled to his own opinion. Everything man needs to earn an honest living is found in the Scriptures.

It has been said that an uneducated man will steal coal off coal trains, but give man an education, and he will steal the railroad. We have seen this taking place in reality in the past months. It is corrupted knowledge that motivates a man to steal billions of dollars from thousands of people.

---

1. James M. Boice, "The Logos" *The Logos* 3 (1996)

THE BEAUTY OF WORK

Our study today takes us to II Thessalonians 3:12, and one of the most important passages of Scripture in the entire Word of God applying to social or **cultural** matters, as well as individual ethics. Let's read verses 10 through 12 in order to get more of the sense of what is being said here.

"For even when we were with you, this we commanded you, that if any would not work, neither should he eat. For we hear that there are some which walk among you disorderly, working not at all, but are busybodies. Now them that are such we command and exhort by our Lord Jesus Christ, that with quietness they work, and eat their own bread."

This is a New Testament commandment. The Law of Moses is not the only place in the Wordof God where we find command-ments. The New Testament is full of them. If Christ is King, and He is, then we should expect Him to issue commands. We do live under Law, but it is the Law of love. The commandments are no less binding.

The Word "command" is used twice in these three verses. I was raised on this sort of ethic. While I was a lad, it never occurred to me that I should grow up and have other people support me. I knew from day one that I was going to have to take care of myself. Two corner stones are found in the life of every middle class American my age: (1) obey the Law (meaning the fellow in uniform wearing a badge), and (2) get a job. What kind of a job it was would depend on what I was capable of doing, but whatever it was - I was goin g to have to get a job. I didn't particularly want to get a job, but I was afraid to face a frowning world without a few dollars in my pocket, and the only way I could put a few dollars in my pocket was to get a job. My granddaddy Kennedy had been blinded when he was a young man, and so he had to push a broom in the high school hallways, and do whatever else he could do to make some semblance of a living. My grandmother Kennedy was an Irish washerwoman, and I am proud to be the grandson of such a couple. They were noble people. I never knew anybody in my young life who expected somebody else to pay his way. We just didn't think of such a thing. II Thessalonians 3:10-12 was a foundation stone in our culture.

When Jamestown was struggling to survive, a redheaded captain named John Smith knew what the Bible said about the work ethic - and consequently, about character. The nobles who came over with the early settlers didn't want to do anything but chase pretty Indian girls through the woods, and stroll on the beach looking for gold. Folks were in danger of starving to death. There was no time or place for liberals with ill-formed notions and who couldn't bear to make folks work. Captain John Smith opened his old Bible and read the riot act to these loafers. They went to work, and the colony survived.

People faced with starvation will go to work or they will steal. They certainly won't starve if there's food around. People with character and Faith in the Bible are the best foundation for social interaction. They are not cruel, heartless people. God is not cruel and heartless. What I am saying is based on sound doctrine which must be followed. We give five dollar bills away to people standing on street comers claiming to be homeless. We hurt for people who have needs. We also hurt for people who are struggling to keep their heads above the water while their government cuts huge amounts of money out of their hard earned wages to give to people who would only take a job as the manager of the country club.

Work is honorable. My son, Jim, is the superintendent of the Southside Water Department. He is the only superintendent the department has ever had. A store is being built here in Southside, and a water line ditch had to be strengthened in front of it. Jim could sit in his office with a white shirt and tie on, and give orders to the other employees. But there are times when it is necessary for him to do some hard work. He always goes to work with his work (ditch work) clothes on. As he was in the ditch in front of the store up to his chest, with shovel in hand moving dirt, a photographer took his picture, and they put the picture in the local paper, with the caption that Jim was an employee of the water board. He is an employee of the water board, but he is also the boss, and he is also willing to get the job done, whatever it takes. He is looking for someone to hire to train to replace him in a few years when he retires, but at this point in time he hasn't found anybody because everybody wants to wear the white shirt and tie (which are also work clothes), and never touch a shovel. Our passage is called "Saint ethics," and what it means is to keep quiet, and get to work.

Used to be, we went to stores operated by the people who owned them. When you went in, you did your trading with the person who owned the store. Nowadays, when you go into a store, usually you do your trading with people who didn't even work there last week, and won't be working there next week, and couldn't care less whether you came in or not. The owners are off at the country club talking about how rich they are, and their place of business is nothing to them but a profit margin, and the customers are nothing but sources of income. America just isn't as sweet as she used to be. She needs the sweetness that people like John Wesley and D. L. Moody brought. She needs the sweetness of school teachers who would lead their class in prayer.

Works has no place in the salvation of a lost person, but it has a great deal to do with the life of a saved person. Good works will not get a sinner any closer to Heaven, but good works will be royally rewarded for the saint. When we are saved, we are created unto good works.

Work comes in different forms. Most people think of cotton fields and white cotton balls when the think of work. Others think of a chopping block and axe when they think of work. Work doesn't always produce sweat, however. There is work that seems easy to some. Work is the production of something of value, whether it be dressed in an expensive suit, or in overalls.

Work - labor - is noble. When God created man, He created him to work. Genesis 1:26 explains God's purpose for man. Man was to have dominion over the earth and all its creatures. That is work. In Genesis 2:15 God "...*put him into the garden to dress it and to keep it.*" Adam was created to be a farmer, a husbandman. Before man was created, God said in Genesis 1:5 that "...*there was no man to till the ground*." and so God created man to till the ground. After the fall, God in mercy gave man work to do because in a fallen state, it would be far better for man to be concerned about his daily bread than to be idle. God never intended man to be idle.

When we get to Heaven there will be work to do. Micah 4:4 gives us the promise that every man will sit under his own vine and fig tree. Most of oursongs about Heaven are pretty shallow. Who would want to walk on golden streets forever? I'd rather feel good rich soil under my feet as I raise luscious vegetables. I love to sing, but I don't want to sing forever. After a few million years of doing what the songs extol and, I assure you, we will be ready to go back to chopping cotton or milking cows. Work is the real blessing.

For now, we must recognize the blessing of work, and seek it out, and earn our own bread with no complaining.

"Whether therefore ye eat, or drink, or whatsoever ye do, do all to the glory of God." "Let him that stole steal no more: but rather let him labour, working with his hands the thing which is good, that he may have to give to him that needeth."

Even the dumbest people can have common sense. Even the most foolish can have common sense. Education doesn't always produce common sense. Common sense tells us many things we need to know to live successful lives. Before there was a school house, people knew there was something noble about

GOOD CREATURES

J ust about every creature on Earth has some human who eats him, or would eat him if he could catch him. I saw the program about the flies that hatch out in the big lake in Africa and come up out of the water in clouds. People along the shore of that lake eat those flies. We have all seen films or pictures of people eating grub worms and other creepy crawlers out of the ground or rotten logs. Rattlesnake and alligator tail are delicacies to some folks, and others consider dog to be culinary delight.

If one human digestive system can process something, all human digestive systems can process it. The reason one person won't eat what another person eats is just because of **cultural** hang-ups. Some folks eat lizards, monkeys, and all sorts of sea creatures. What one human stomach can digest, allother human digestive systems could digest if they had the same health and cultural conditioning. This is not to say that people ought to eat anything they can get their hands on, it is simply to say that they could if they had to.

God did not create animals for food, but He did say that they were "good." Not good to eat, but good. Genesis 1:20 and 21 tells us about the creation of the animals, and then it says, *"...and God saw that it was good."* The animal creation was good when God created it, and could have been eaten if that had been God's plan for them. The "beasts of the field" were probably the domesticated animals, sheep, cows, chickens, etc., which man could have eaten. *"For every creature of God is good, and nothing to be refused, if it be received with thanksgiving.* (I Timothy 4:4).

It is not what you eat so much as how you receive it. There should be no mystery about having a happy stomach. Food must be blessed before it is consumed. With God's blessing upon it, all food is good. I remember my daddy "returning thanks." At every meal we would all bow our heads and he would say, "Our Father, we thank thee for these foods and nourishments. For Christ's sake, Amen." Those were sweet moments for me because I loved to hear my daddy pray. My mother was a good cook. Once my daddy shot a ground hog in Mr. Smith's corn patch beside the north fork of the Holston River over near at a spot we called the "Big Hole" close to Mendota, Virginia. He insisted that we bring it home so Mother could cook it up with some yellow sweet potatoes. He remembered I Timothy 4:4,

"For every creature of God is good, and nothing to be refused if it be received with thanksgiving."

I remember my mother gagging as she carried a big roasting pan full of groundhog oil out of the house. It was pretty gross. People who know more about cooking ground hog could make it a delicacy, no doubt people on Earth eat just about everything that moves, and a great number of things that don't

The Lord tells us in verse 3 this same thing. He speaks of,

"...meats, which God hath created to be received with thanksgiving of them which believe, and know the truth."

I frequently go to homes of folks to eat This is an honor, of course, and I always eat what is set before me with thanksgiving, and without asking nosy questions. If I must eat blood pudding, I don't want to know it When a woman works her fingers to the bone all day to set a table, a fellow ought to eat it like he enjoys it After we pray and thank God for what's on the board, how can we then complain about it? We must be careful about that Verse 5 tells us that any creature *"...is sanctified by the Word of God and prayer."* All a ground hog needs to make it good to eat is the Word of God read, and prayer offered for it. ('Course it doesn't hurt if it is cooked right.)

Animals were not created to be eaten. Nothing was to be eaten at the beginning except plants which were created for the purpose of converting the energy of sunlight into food for all other creatures . We do not know what sort of teeth humans had at that time, but we suspect they could have had what we call predator canine teeth which could have eaten flesh. The panda in China eats nothing but bamboo, yet it has what appear to be predator teeth. Tyrannosaurus rex, the big dinosaur that was supposed to be the king of all predators, had teeth that grew on his gums. His teeth were not embedded in his jaws like a predator's teeth must be.

Man and animals did not eat flesh until after the great flood. That reference is Genesis 9:2-4.

At this point in time, God had not yet said that

"... the life of the flesh is in the blood: and I have given it to you upon the altar to make an atonement for your souls.",

But it would come about 800 years later as God continued to unfold His plan for man's redemption. Meat is now a part of a healthy diet, and we are permitted to eat anything as long as we thank God for it. The Lord's desire is that we live a healthy and productive life. Our culture should not dictate what we can eat, and what we cannot. God has made all things good.

In the new environment man would have to live in after the flood, it would have been impossible for man to survive by natural means without being allowed to eat meat. God **curse**d the plants so that they would not yield to man the necessary nourishment for his survival, and so man would have to receive part of his nutrients from animals which could eat the plants, and convert plant energy into food for humans. Plant energy is simply stored sunlight. Man is the final user of this stored sunlight. When someone takes a bite of hamburger, he is eating a great deal of plant matter that the cattle had eaten and the plant matter was stored energy from the sun.

The reason this verse, I Timothy 4:4 is written in the Bible is because there were religions that were " ...***commanding to abstain from meats,***" through the authority of their priests. One of the worst elements of the Law of Moses was the hierarchical priesthood. ("hierarchical means ranks of authority). The Law was a curse, and one of the main reasons was because of the priesthood. The wickedness of priests is seen in the evil character of Hofni and Phinehas, the sons of Eli, in I Samuel 2:12-17 and I Samuel 3:13. Even the sons of Moses were evil individuals whom God was forced by their wickedness to execute. The Catholic priesthood through the ages has been a sordid story of sodomy, lust, and lies. Some of the worst men in history have ruled from the papal throne. Read Leviticus 10:1,2. Read *Halley's Handbook* for a brief description of the priesthood, pages 874 through 899.[1]

In this passage in I Timothy, the Lord lists five very serious spiritual crimes of the persons spoken of in verse 1: (1) they departed from the Faith; (2) they gave heed to seducing spirits; (3) they spoke lies in hypocrisy; (4) they forbade to marry; (5) they commanded to abstain from meats. Some men have such a lust for power that they will join forces with the very devil himself, and take from others some of life's sweetest blessings, even their very souls. Noticehow these sins build one upon another; departing from the Faith leads one to giveheed to seducing spirits; heeding seducing spirits causes one to be a liar; and lying leads to making such anti-scriptural commands as to what a person can eat. Man cannot improve upon what God has given us. Nothing but harm can result from adding or taking away from, or diluting the Word of God. All sins build upon each other. No sin stands alone. All sins injure more than one person.

I am a priest. I am a priest after the order of Melchisedek. The priesthood to which I belong is not an hierarchical priesthood. No priest in my priesthood is above another. My thirteen year old granddaughter has the same authority before the Throne of Grace as I do. My pastor has no more authority as a priest than I have. We are all equal in Christ's priesthood. Christ is the High Priest of our priesthood. He is a member of the order of Melchisedek.

Everything God created was good. God created everything to live in beauty, harmony, and tranquility, in fellowship with Him. How much better is this belief than to believe that there is no God, or that He is not limitless in power, knowledge, and duration. If the creation story were not the true account of the beginning of all things, it would still be the best thing for humankind to believe. If I were an atheist, I would want everyone else to believe in God.

God created the animals good. I think animals are still good, even under the **curse**. I love all creatures of the earth, and enjoy them immensely. I even love my neighbor's dog! I am sorry that they suffer from the results of the curse, and I am glad that one day God will restore them again to their original estate.

---

1. Henry H. Halley *Bible Handbook* (Chicago 90, Illinois: Henry H. Halley, 1957), pps. 874-899

GRACE IN ETERNITY

Who hath saved us, and called us with an holy calling, not according to our works, but according to his own purpose and grace, which was given us in Christ Jesus before the world began."

The thought of eternity is a foreboding idea because we cannot comprehend it, and because God dwells there. Thinking of eternity is like thinking of a different world, or a different dimension and we feel like aliens adrift in space. "Everlasting" means never ending in the future, and that is the sort of life we humans have. Our body must undergo a change we call death and resurrection, but we, that is our soul and spirit, will never die. We who are saved have everlasting life, and those who are not saved have everlasting death.

There are questions about the exact second when a person dies. With the present advanced state of medical science, it is sometimes difficult to know when a person is dead. Some people are said to be brain dead, and others linger so that questions arise about when they actually died. We do not need to be concerned about those problems because God knows exactly when to beckon the soul and spirit out of the body. Medical personnel do have to be concerned about that.

There is another sense in which we do see death, for old Simeon in the temple had been promised by the Lord that he would not see death until he had seen the Lord Christ. That is because "*...it is appointed unto men once to die, but after this the judgment.*" Our body must die, and we must deal with that truth all of our life. Death is a part of living, the last part of living, for our body, but after death the redeemed will never have to deal with it again. Those who are lost must deal with it for eternity, for death and Hell will be cast into the lake of fire with them, according to Revelation 20:14. There is nothing that the human mind can imagine that is worth risking such a fate for. Supposing it is true, sinner friend? Supposing you are wrong in what you believe? Imagine the stark horror of waking up in Hell; to suddenly realize that everything the Bible says about Hell is true?

About the time of the War Between the States, Elizabeth Clephane wrote, "Beneath the cross of Jesus I fain would take my stand, The shadow of a mighty rock Within a weary land; A home within the wilderness, a rest upon the way, From the burning of the noontide heat and the burden of the day."

According to II Timothy 1:9, God's purpose and grace were given to us before the world began. Here is what that verse says,

"Who hath saved us, and called us with an holy calling, not according to our works, but according to his own purpose and grace, which was given us in Christ Jesus before the worldbegan."

The world began. Genesis 1:1 tells us that "...*In the beginning God created the Heaven and the earth.*" It is not difficult to know when the beginning was. When God created the Heaven and the earth, that was the beginning of time, and the beginning of that part of eternity in which we humans exist. John 3:15 says,

"That whosoever believeth in him should not perish, but have eternal life", and John3:16 says, *"For God so loved the world that he gave his only begotten Son, that whosoever believeth in him should not perish, but have everlasting life."*

In verse 15 we receive the eternal life which is the life of Christ, and in verse 16 we have everlasting life which is the life we have as humans which begins when we receive Christ Jesus as our Savior.

The unsaved live forever, but in a state of death. That state of death is separation from Jesus Christ who is life. Death is separation from life, and so separation from Jesus Christ is a form of death. John 5:29 tells us that all

". ."shall come forth; they that have done good, unto the resurrection of life; and they that have done evil, unto the resurrection of damnation."

I am a saved sinner. As such I have everlasting life, and now live in eternity. As a saved sinner I have committed many a sin. I was saved when I was but a lad, and over the years, I have done many things which have broken my Heavenly Father's heart. I have given Satan much to accuse me of before my Father's throne. I'd give anything if it were not so. If I could live my life over, I would try harder to avoid taking a sip from any cup of iniquity. What my sins are is not for public display, for it is of no glory to the Lord for the world to know what I have done wrong. "My sin - oh, the bliss of this glorious tho't - My sin - not in part, but the whole - Is nailed to the cross and I bear it no more, Praise the Lord, praise the Lord, O my soul! It is well with my soul, It is well, it is well with my soul." P. B. Bliss had it right when he wrote the soul-stirring Words in 1873, for a soul in Christ can always perceive the caring presence of the Holy Sprit. H. G. Spafford wrote the beautiful music.

Am I still saved after sixty-two years of sin and failure? My Russian Saint friend that came here to visit me would say that I am not saved, because he believes that if you sin after you are saved, you are lost again. You may object that a person who believes that idea is not saved, and I would not debate that with you at all. This man got in a rage because some young people at a church we visited had made the statement that they could do anything they wanted to do, and he did not understand what they were saying. I think one of them must have smoked a cigarette while they were away from the church, because he was saying that smoking cigarettes would send you to Hell. Please believe me when I assert that I condemn that sort of thing, and God does too, but He would never condemn a human soul to Hell for that. If He would, then I would be lost, too, because I have done things worse than smoking a cigarette, even though I have never smoked a cigarette. My Russian friend has done worse.

Why is it that a saved person's sins cannot condemn him to Hell? There are many reasons. First of all, we are not saved by being good, we are saved by relying on the grace of God for forgiveness through death, burial, and resurrection of Jesus Christ. But please note what our verse says. II Timothy 1:9 says,

"Who hath saved us, and called us with an holy calling, not according to our works, but according to his own purpose and grace, which was given us in Christ Jesus before the world began."

The Lord says here that we were recipients of His purpose and grace before the world began. How can a thing done in eternity be disannulled by something done in time? How can a thing done before we actually existed be disannulled by something we do in our lifetime? A thing given us in eternity is an eternal thing. God saved us and called us with an holy calling before we had produced any works, good or bad. A sinner can only go to Hell because he refuses to accept the salvation God provided for him before he was born. After he accepts it, he can never lose it. He would certainly never reject it.

People go to Hell because they do not accept the salvation God has provided for them. Your salvation, completely paid for and free of obligation, is waiting for you to claim it. You can claim it wherever you are.

The dictionary defines "eternity" thus: "1; The totality of time without beginning or end. 2. The state or quality of being eternal. 3. Afterlife; immortality." To many of us eternity occupies a big blank in our mind. We tend to think of eternity as floating in space, like an astronaut who has lost his life line to the space shuttle. Only God can explain eternity, and He elects not to in His Word. We can get closer to explaining eternity by believing in the eternal LORD of time and eternity. We know, though we cannot explain, that God is eternal, and His existence explains eternity to our simple minds a reality. God is perfect and therefore never changes. He was a God of grace to Abel the same as He is the God of grace to us. Since grace is an attribute of God, then it must be eternal.

Saints receive God's grace, as well as His purpose when we are saved. His grace and purpose have been ours in Christ ever since Christ was crucified before the foundation of the world (Revelation 3:8). None of the prophets even existed when God's Word was perfected in His thoughts. The devil did not exist in eternity because God created him, but his story was in the unwritten Bible. His Word is settled in Heaven. It cannot be changed or altered, though men have tried. I believe the A.V. is the Bible God gave man, but I do not believe it is the Word settled in Heaven. Grace is that wonderful, indescribable quality of God that made man possible in the first place, and saved in the second place, and kept for eternity in the third place. Most of what men write is trite, and much of it erroneous, but God in His mercy and grace blesses the truth we write, and uses it for His everlasting glory.

GOD'S WORK

If the Constitution of the United States were written in such a way thatthe average American could not read and understand it, the freedoms it guarantees would not be of any value to us. If we had to rely upon the constitutional Lawyers and judges to tell us what the constitution means, our freedoms would be what the constitutional Lawyers and judges say they are. Unfortunately, that is what we have come to at this point in our history, and I realize that the business of the **Supreme Court** is to interpret the Constitution. But I stand by my assertion that the average person can read and understand the Constitution well enough to know pretty well what his freedoms and responsibilities are. If we must depend upon somebody to tell us what the Bible means, then our Faith must rest in that person.

I was in Siberia when the Russians were voting on their new constitution. I pleaded with the people to go vote, but most of them felt it was of no use. One Russian businessman told me that it was just a piece of paper that meant nothing. It is too bad that is more true than we would like to think.

The Bible is difficult to understand in some places, and must be interpreted or explained by mature and wise teachers. The Word "interpret" is found in two instances in the Bible. First of all, **Joseph** was called in to interpret Pharaoh's dream, and we all know the result of that. Secondly, the Word is used by Paul in I Corinthians where he is discussing tongues, and the conclusion Paul reaches in that discussion is that if there is no one around to interpret what a person says in an unknown tongue, the person should not speak in an unknown tongue - period.

Everyone from time to time needs an explanation of a difficult passage. It is not something to be embarrassed about. I have needed difficult passages explained to me more than once. It has been truthfully said that the Bible is its own best commentary. Since it is so interconnected, there is nearly always a passage somewhere else that will shed light upon the hard passage. Since the Lord is a God of compassion and reason, we can be sure that we can always discern the perfect will of God for our particular talents and gifts.

For several days we have been studying passages in the New Testament that refer to passages in the first three chapters of Genesis. I have done this because of the importance of the first three chapters of Genesis. Satan is shrewd enough to know that if he can weaken man's Faith in this part of the Bible, he can destroy a great number of souls. It is time to point out more in detail how Satan attacks these chapters in particular. Satan has a special weapon he can turn upon these chapters. It is called evolution. It is a special weapon because evolution can be mixed with science in such a way that unthinking people become confused, and cannot discern between the science and the philosophy. The first eleven chapters, of Genesis are in a real sense scientific because they involve the work of God in creating and shaping the earth, and that work is what established matter and the Laws that govern it; and that is the stuff that science is made of.

These attacks on Genesis can be classified under three very broad headings: 1. several people were involved in writing Genesis, and therefore Moses did not write all of it and so it was not inspired; 2. Genesis is just mythology; 3. Genesis was written long after Moses died.

In the first place, it would not matter greatly if many people were involved in writing Genesis. In fact, some fundamental scholars believe that Adam wrote books, and that Moses had these books in hand when he wrote Genesis. Seth and many other Genesis chapter 5 people could have, and probably did, write books. Many people were involved in writing the Bible. There is reasonable internal evidence to lead us to believe that Genesis was written by several people, but whoever wrote it, did so under the divine inspiration of the Lord, and Moses may have used that when he wrote the book. He may have compiled many records into the final book. Approximately forty men over a period of seventeen centuries wrote the Bible.

The Bible tells us that "...*holy men of God spake as they were moved by Holy Ghost.* ", and so it doesn't matter who penned the Bible, (except that they had to be holy men of God), because it was God the Holy Ghost Who is the primary Author. There is the living Word and the written Word. Both are perfectly compatible. The written Word is a red carpet to the living Word. The living Word is the manifestation of the written Word.

In the second place, Genesis is not mythology, nor is it derived from ancient mythology. There is no scientific evidence to support such a claim. Archaeologists have never found anything that would indicate that Genesis 1 and2 are not actual, true accounts of two real people who were created perfect by an omnipotent God. Unbelievers are entitled under the constitution to their opinions, but not before God.

Dr. Isaac Asimov was well acquainted with the Bible as a secular book. He points out that "...millions of people today know of Nebuchadnezzar, and have never heard of Pericles, simply because Nebuchadnezzar is mentioned prominently in the Bible and Pericles is never mentioned at all. Millions know of Ahasuerus as a Persian king who married Esther, even though there is no record of such an event outside the Bible. He goes on to say, ". Biblical interest was centered primarily on developments that impinged upon those dwelling in Canaan, a small section of Asia bordering on the Mediterranean Sea. This area makes only a small mark on the history of **civilization** (from the secular viewpoint) and modem histories, in contrast to *the* Bible, give it comparativelylittle space."[1]

Dr. Asimov's point that the Bible concerns important things that are not mentioned outside the Bible simplv is right. God planned to keep His book from being subject to any sort of scientific proof or disproof. The Bible cannot be proven to be the Word of God by scientific methods, and neither can it be proven not to be the Word of God by scientific methods. The Bible must be received by Faith.

It is very important to note that never has there been any evidence of any sort from archaeology or any other science that has shown the Bible to be wrong or unreliable. All of the scientific evidence available to man today tends to show that the Bible is right in its account of creation. None of the evidence shows that the Bible is wrong.

In the third place, the notion that the Bible was written a thousand years after Moses died is without foundation in fact or truth. This notion arises from the preconception of skeptics that man evolved, and so couldn't know how to write by the time of Moses.

Dr. Morris quotes Ralph Linton a well-known evolutionary anthropologist as saying, "Writing appears almost simultaneously some 5000-6000 years ago in Egypt, Mesopotamia, and the Indus Valley." The Scripture teaches that man was created perfect (Genesis 1:31), and therefore the first people had to be the most intelligent people who ever lived. To think that they could not have learned to write in a very short time is foolish. Evolution teaches that man began as a tiny thing in a puddle, and so didn't know anything. That's mythology.

And so no one can believe that Genesis 1 - 3 are not an integral part of the Scriptures, equal in every way to all other parts of the Scriptures, and still believe the New Testament, for the New Testament is rich in references to those foundational chapters. That is the purpose of our study through the New Testament, seeking out some of those verses that pertain to Genesis 1 - 3. The Bible was written for every man, but every man was not given the authority to stand in judgment on it. Man was given the authority to rightly divide it, but man was not given the authority to chop it to pieces.

When the Bible became no more than "just a translation," it lost most of its authority. In a world that hates God because they don't understand God, it is necessary to have an authoritative voice. Saints are few who give a good daily testimony of God. The Bible does need some editing. It was edited several times between its original translation in 1611, and the final edition in 1769 when it became the A.V. Another editing now would be a good thing, but the main text should be left intact. When God is ready for His Word to be fixed, He will see that it is done according to His divine purpose.

1. Isaac Asimov, *Asimov 's **Guide to the Bible*** (Doubleday Company: Garden City, New York), 1969, pp. 9,10.

2. Henry M. Morris *The Genesis Record* (Baker Book House: Grand Rapids, Michigan: 1976), p. 24

HOPE OF ETERNAL LIFE

P aul said, *"In hope of eternal life, which God that cannot lie, promised before the world began;"*.

God promised eternal life. Who else could make such a promise? Dr. Isaac Asimov, of whom we spoke on the last broadcast said, "Although the time of death is approaching for me, I am not afraid of dying and going to Hell, or (what would be considerably worse) going to the popularized version of Heaven. I expect death to be nothingness and by removing from me all possible fears of death, I am thankful to atheism."[1]

The possibility of being wrong is something every man ought to consider, no matter how great is his intellect or his knowledge of a subject. I have been wrong when I would have staked my life that I was right. If a man is wrong about something worldly, he can nearly always make amends for it. But if a man is possibly wrong about a matter that he cannot change forever, he is very foolish if he does not try to make that mistake right before he dies.

The matter of going to Heaven is something that no man should take a chance on. Going to Heaven, of course, is only a matter of belief, no one can prove he is going to Heaven. I know I am going to Heaven, but my knowing that is not going to be accepted by unbelievers. The unbeliever never seems to consider the possibility, no matter how slight that he is wrong in his conclusion that there is no God.

Titus 1:2 has a phrase in it just like the phrase in II Timothy that weconsidered in the last essay. It speaks of something being done for us before the world began. Listen to the verse,

"In hope of eternal life, which God, that cannot lie, promised before the world began."

Who was eternal life promised to? Did you ever promise something to one of your children before he was born? That is altogether possible, you know. You can promise something to someone before you meet them, or before they arrive where you are. A person does not have to be present for you to promise them something. Again, we see that our salvation was a matter occupying the mind of God from eternity past. And it is a matter which was sealed in eternity, not to be loosed or broken in time.

The hope of eternal life is not to be taken lightly. The alternative to eternal life in not sleep in the sweet dark earth, as Dr. Asimov and others liked to think. Corliss Lamont is another atheist who deceived himself into thinking that life was no more than the beat of a four chambered material organ inside his chest. Here is what Lamont said, "If human beings are to be happy and to enjoy life, it must always be during some period of time describable as *now*. What the future-worshippers do is to ask each succeeding generation to sacrifice itself in working exclusively on behalf of a distant Utopia that may or may not come. The humanist asserts that, from the viewpoint of human happiness and the sum total of good, today is just as significant as tomorrow and the current year just as significant as any a decade hence."[2]

But how can a human be happy in the everlasting *now* if he has no hope for the future? How can one be happy *now* if he is uncertain about the future? For a person to be completely happy, there must be a measure of confidence in the security of tomorrow. There must be hope, and that is what God promised us before the world began. Humanists and atheists take comfort in annihilation. The **darkness** of the grave is a solace to them. They have so utterly eliminated God from the cosmos that they place their hope in nothingness.

What is worse than a false hope? Have you ever been disappointed? It is miserable to have a bright hope shattered. It is especially depressing to have a hope shattered suddenly and without remedy. You may be suffering right now from such a calamity. I certainly hope not, but if so, I pray that you will soon be able to shake it off. How awful to have a precious child in whom you place great hope for his future, only to stand helplessly by while he totally wrecks his life. What an awful burden on parents! But yet, in the deepest despair there is hope. As long as that wayward one lives, his very life engenders hope in the heart.

The hope of eternal life is like a precious disinfectant. People who believe they will live again develop an inner purity that blesses all about them, while those who look forward only to death and dirt tend to carnality and sin.

Paul wrote in I Corinthians 15:19, *"If in this life only we have hope in Christ, we are of all men most miserable."* If our hope in Christ is a false hope, then we are of all men most miserable. We have sacrificed our money to pay the way of missionaries while others spent their money on themselves. We have worn ourselves out doing honest day's work while others loafed and hid from the boss. We have allowed ourselves to be robbed by a voracious government while others falsified their tax returns. We have been slapped viciously when we turned the other cheek. Oh, if Christ be not risen, how miserable we are!

But if Christ is risen, and He is! then how miserable are those who are ignoring Him. People in our world ignore the worship of God, choosing instead sporting events and other events that the world offers. These activities cannot satisfy the hunger of the soul for fellowship with the Creator.

Hope is a part of our divine inheritance. God cannot lie. He gave it to usbefore the world began. Hope had to be an element of human existence. Even Adam and Eve must have experienced hope, for they were to plant gardens and eat of the fruit. They were to have children, and there cannot be a child without hope for the child. Hope has to do with things not seen. Hope is the bright evidence of things not seen, just as Faith is. Faith and hope are not synonymous, but there can be no substantial hope without Faith. The ungodly place their Faith in luck or circumstances, for even the man who hopes that death will bring annihilation hopes that he is right, and believes in his ability to make right choices. He hopes that his life will last a long time.

I wonder if animals have hope. Does a deer strolling through the forest hope that he will not get shot that day? Does he hope that he will have plenty of acorns next year? Romans 8:19 tells us that animals have an earnest expectation, which I am sure would be about the same as hope. Could this aspect of creatures have evolved?

Evolutionists cannot begin to explain or understand the consequential matters of existence. As with all matters that count for time and eternity, we must go to the Word of God for the answers. Dead matter could not evolve such a phenomenon as hope.

The Gospel of Jesus Christ is so precious because it holds out a hope that is steadfast and sure. I love the Gospel because it reaches out to people, and brings a peace that passeth understanding which is grounded in the blessed hope. We shall die. What a depressing thought! To give up life and all its bright blessings, and to die. Indeed, death would hold no fears if that were truly the end, as the humanist believes, but who can know for sure? Who is so foolish as to allow that event to arrive without making preparations for it? Hope will not save. Hope will not make it so.

In another verse that strangely enough tells us that God cannot lie, we read of a hope set before us. Hebrews 6:18,

"That by two immutable things, in which it was impossible for God to lie, we might have a strong consolation, who have fled for refuge to lay hold upon the hope set before us."

That is very significant to me, that we are assured of our hope because God cannot lie. We have our hope fixed in an honest, eternal God of integrity.

"While there is life, there is hope – never was there a truer Word. Do not, believe they will live again develop an inner purity that blesses all about them, while those who look forward only to death and dirt tend to carnality and sin. Paul wrote in I Corinthians 15:19, *"If in this life only we have hope in Christ, we are of all men most miserable."* If our hope in Christ is a false hope, then we are of all men most miserable. We have sacrificed our money to pay the way of missionaries while others spent their money on themselves.

We have worn ourselves out doing honest day's work while others loafed and hid from the boss. We have allowed ourselves to be robbed by a voracious government while others falsified their tax returns. We have been slapped viciously when we turned the other cheek. Oh, if Christ be not risen, how miserable we are!

But if Christ is risen, and He is! then how miserable are those who are ignoring Him. People in our world ignore the worship of God, choosing instead sporting events and other events that the world offers. These activities cannot satisfy the hunger of the soul for fellowship with the Creator.

Hope is a part of our divine inheritance. God cannot lie. He gave it to us before the world began. Hope had to be an element of human existence. Even Adam and Eve must have experienced hope, for they were to plant gardens and eat of the fruit. They were to have children, and there cannot be a child without hope for the child. Hope has to do with things not seen. Hope is the bright evidence of things not seen, just as Faith is. Faith and hope are not synonymous, but there can be no substantial hope without Faith.

To Darwin

Sagan, Dawkins, Al Capone;
of all these crooks,
he stands alone:
wrote heinous books
to steal men's souls.

Error, deceit, and lies
to black man's mind,
till he hopeless dies –
for eternity blind
for he lost his soul.

'Neath the Abbey floor interred;
blest soil of saints.
Newton's bones were stirred.
Such wickedness taints.
But not his blessed soul.

Is God a fact or fiction?
Crooks assert the latter,
this causes lots of friction,
and that is what's the matter?
And down go men's souls.

He could have stopped his birth,
and killed him in the womb,
but God allowed him to come forth,
men's precious souls to doom,
or give to Him their soul.

GOD'S REST

Hebrews is a grand book, and not as well understood as it deserves to be, I am afraid. Written to Hebrew Saints by an unknown writer, it speaks of the priesthood of Jesus, and is full of the perfume of the Old Testament. Hebrews has at least eight references to the first three chapters of Genesis, and makes it clear, as does John and other writers in the New Testament, that it was the pre-incarnate Christ Who did the actual work of creation. Hebrews 1:2 says that God

"Hath in these last days spoken unto us by his Son, whom he hath appointed heir of all things, by whom also he made the worlds." Just as *"... God was in Christ, reconciling the world unto himself..."* as we are told in II Corinthians 5:19, so he was in Christ, creating all things.

But we shall not use this great verse for our study today, but rather, we will go to Hebrews 4:10, and read,

"For he that is entered into his rest, he also hath ceased from his own works, as God did from his." Genesis 2:2 tells us that *"...God blessed the seventh day, and sanctified it: because that in it he ended his work which he had made: and he rested on the seventh day from all his work which he had made."*

On the **calendar**s of the world, we find seven day weeks. Time is regulated by the movement of the sun and moon, but the seven day week was established upon the truth that God created the cosmos in six days, and rested on the seventh day. In Russia where so many proclaim loud and clear that there is no God, the calendar is composed of seven days. The only difference in theirs and ours is that ours begins on Sunday while theirs begins on Monday. Atheists are called liars by their calendars.

God created the universe in six days, rested on the seventh, and probably while He was resting, Adam and Eve sinned in the garden, and on the eighth day, the Lord God, allowing grace to override justice, went back to work to redeemthe human race, and He has not rested from that labor to this day, and will not until the new Heaven and the new Earth are completed. God will not rest while there are people lost and dying, and while His own children are engaged in mortal combat with their adversary, the devil.

It is true that Christ has finished the redemptive work that brought Him to the earth, but He is not resting today because He is seated at the Father's right hand, making intercession for the saints. The Holy Spirit has finished the work of producing the Holy Scriptures, but He is not resting, He is busy convincing sinners that they must accept Jesus Christ as Savior. Nor must the church rest while there is a lost one wandering in the shades of night. "Work," commands the old hymn, "for the night is coming when man's work is done." You and I must work, for there are souls for which we are responsible, and they must be warned and taught. There will be a day of rest as verse 9 tells us, but it waits over the horizon in another day.

Honoring the seventh day by going to church and then straight home can be a great testimony to the world that we truly believe that God created the universe in six days and rested on the seventh. It is no more legalism to honor the seventh day than it is legalism to refrain from stealing, adultery, or murder.

Only an emancipated slave can tell you what liberty is. A man born free knows not the true meaning of liberty. The Saint understands liberty, for he is a liberated slave while the sinner is still a slave and cannot know liberty at all. Allthe liberties the sinner boasts of are but chains on his soul.

My salvation is complete, perfected and consummated because I have received Jesus Christ as my Savior, and so as far as that is concerned I am at perfect rest and peace. I can enjoy my life because I am not in religious bondage to anybody or anything. I am not forced to do inconvenient, humiliating things to please some god or priest. My soul is at rest in the knowledge that, like Paul,

"...I know whom I have believed, and am persuaded that he is able to keep that which I have committed unto him against that day." (II Timothy 1:12).

- 250 -

My soul and spirit enjoy a sweet repose in the arms of the omnipotent Jesus. I am free to labor in the fields that are white unto harvest.

God's work of creation was perfect. He completed everything. His sabbath rest was perfect, and so is my rest perfect because He did my redemptive work. But the Lord did not get to rest long. He had to begin the process of redeeming lost humanity. God has not rested since Eve bit into the forbidden fruit. Since His resurrection Christ has been at God's right hand making intercession for us.

Yet I am not at perfect rest altogether, for I am not home yet. There remains a perfect rest which will include my body. I am weary and old. I find I need more rest all the time, but my work makes demands on me. Sometimes I fall asleep at my computer and wake up with my neck hurting like a toothache. There will be a perfect rest. Yet my rest now is superior to that of folks who have not God's rest.

While searching the Internet for a certain subject I ran across a book written by an abortionist who performed illegal abortions back in the 1930's. This man's story is not only an example of how human compassion can lead one into evil behavior, but it reveals how age becomes unhappy for people who do not have God's rest.

This man begins his book by writing, "Sometimes I find myself thinking wistful of the days when I was young and sure of myself and my future, when I thought the solid ground under my feet was a foundation for an air castle, and when right and wrong were very definite things, and black was black and white was white and I would have nothing to do with the gray."[1] He then proceeds to relate the tale of how he went from a highly ethical young doctor to an illegal abortionist. A girl came to him begging him to get her out of trouble. He felt so sorry for her that he performed the abortion, and that began his career.

His statement motivated me to stop and think about my own life, nearly all of which is behind me. What was I when I was a young and fiery preacher, and what am I now, and what has the intervening years done to me? Was I sure of myself and my future? Can I think objectively of this, or will I romanticize it? I am human, and therefore I know that my memories will be romanticized, but I did my best to think soberly about it. Can I be perfectly honest, or will I say what I think will be the best testimony for a preacher?

The truth is that I can't remember my youth well enough to analyze it in any detail. I can state emphatically that I have never doubted that I was saved when I was six and a half years old. I do remember that as a young pastor I had no doubts about what I believed. I was thoroughly, solidly, unapologetically fundamental in my Bible beliefs.

All of that, though, is not the important part. It is all gone and can never be changed one way or another. What am I today? That is of supreme importance. The abortionist was leaning upon what he was in youth for comfort in his old age. He was appealing the healthy attributes of his youth for redemption from the sins of his mature years. Some people can find solace in the memories of their youth when they felt young and pure and virtuous, but that solace always falls short of perfect rest. There is always an element of regret if nothing else. If one must remember himself virtuous in his youth, mourning the loss of it, he cannot be completely at rest. God, the supreme psychologist declares that

"...the wicked are like the troubled sea, when it cannot rest, whose waters cast up mire and dirt." (Isaiah 57:20).

Yet I too, must confess that I am not at perfect rest because I remember the sins of my youth, and as **David** cried, *"...my sin is ever before me."* (Psalm 51:3), so I groan when I think of some of the things in my past.. A man who is not troubled by the gross sins of his youth is a man of little integrity.

So how is my old age and its store of memories any better than the abortionist? I have been forgiven for every sin, and I have the hope of everlasting life. That is perfect soul rest.

---

1. Martin Avery, (nom de plume) *Confessions of An Abortionist* (Girard, Kansas: Haldeman-Julius Company, 1939).

DOWNWARD CHANGE

*A*nd saying, Where is the promise of his coming? for since the fathers fell asleep, all things continue as they were from the beginning of the creation."

The promise of His coming is the believer's blessed hope. His appearance in the clouds is a sight I long for. Oh, that it were today

Through the centuries Americans have worked with a dream in mind and heart, and that dream was that their children would be better off than their parents. This is the American dream. The dream of owning one's own home, and raising a family; raising the children to build upon the work of previous generations is the American dream. That was the dream of my parents, and the dream of my wife and I as we raised our sons. But now false religions and philosophies have infiltrated our culture, and brought with them a host of problems that is reducing America to mass confusion. Easy divorce and welfare have just about destroyed the American dream, not only for the middle class, but those the middle class has been forced to support.

You would think that with the soon arrival of the third millennia, and a fifth of the way through America's third century, that we would have built a paradise on earth by the year 2000. Such is sadly not the case. There are a great number of good things about America in 1995. There are still many places where people can live in peace and tranquility. There have been medical advances that make human life in America infinitely better than it was in years gone by. We are rich in houses, and cars, and all sorts of electrical things, and good beds, and marvelous kitchens. Indoor toilets are such a comfort that kings of old would have invaded another country for such a luxury. Food is so plentifulthat America has the unique distinction of being a nation that can eatfor fun and throw out garbage by the ton. Americans are rich and spoiled and bored. Have books lost their qualities? The touch of a book transports the reader to worlds unknown. The pages of a book can make one young again – or old. Torrents of books flow from America's presses, yet people get bored. How can it be?

I was terribly distressed the other night when I heard again that suicide among young people in America is one of the leading killers of kids. Childrencommitting suicide! I can hardly believe it. I think of my life as a boy, and how good life was, and I wonder how on Earth a child can bring himself to end his life. My life was not good because we were rich, but the children committingsuicide are not poor. What is wrong? Are there no more trails for kids to hike, and no more streams to fish? Are there no more summer suppers of macaroni and cheese to make living through the day worth the trouble? Are there no more good cowboy shows to go to on Saturday afternoon? Are there no more vacant lots where kids can play ball without being bothered by adults? Are there no more apple trees with stout branches to climb around in? Where is the old swimming hole with sycamore tree diving platform? Are there no more Lawns to mow for a nickel which will buy a cold Nehi? Are there no more Sunday school teachers who love children enough to take them to revival meetings and then out for a milkshake? What has happened to the world that it is not a fit place for children to live in anymore? Why are the church bells silent?

I think of my three sons' children, and the bright, healthy, happy children I see in the churches I visit, and I wonder why all children can't be like them. Are the parents of the world so blind that they cannot see that America can still be a good place to raise children if they would just give God His rightful place in their lives? Sinners make the mistake of thinking that they can raise their children without God's help, and they get along pretty well for seven or eight years, but then something happens, and before those parents realize it, their children are unmanageable. What happens is that the child reaches the age of accountability? When a child becomes a sinner by choice as well as by birth, he is at great risk, and he is very hard to manage. Children can get saved, and live a successful life even in this perverse age, guided by wise parents and devoted Sunday school teachers. But they must get saved. No one has ever lived successfully without being saved, and no one ever will,

"For what shall it profit a man, if he shall gain the whole world, and lose his own soul?"

Sober sinners go to Hell the same as drunk ones, and honest bankers go to Hell the same as the crooked ones. A man must be born again to enter Heaven's gates, no matter what his lot in this life may be.

What a joy it is to be in a church that abounds with children who are submissive to Jesus Christ. I went over into Georgia to a camp meeting, and what exceeding joy I experienced there. A girls' home over in South Carolina came, and the thirty or forty girls in young people I have ever seen. These high school age girls had been through some very bad experiences, but now they are saved. There is a holy glow on their faces. There is a virtue about them that lends a purity of heart to all who come near them. Their testimonies are given with a power of the Holy Spirit that moves the listeners to a greater desire to live righteously. Holy women have enormous spiritual power.

I love a church that puts spiritual matters above the profane for young people. I'm thankful for a church that encourages young people to memorize Scripture. Thank God for youth pastors who take their young people to revivals and other meetings at sister churches when the opportunity presents itself. How precious is a Sunday school teacher who will take interest in his Sunday school class all through the week, and uses imagination to devise ways of encouraging them to live closer to the Lord. I appreciate churches where the young ones are so spiritually inclined that they do not settle just for pizza on Sunday night. Thank God for churches that still have bus ministries, and where the old folks are glad to support such a work. It is good to know that there are churches where the children sit in the auditorium and learn the blessed old hymns of the church, and hear the mighty preaching of the Word of God, and the prayers of the saints. Bless those good churches where evangelists are as welcomed as singers. Thank God, we still have good churches, where sensible parents can depend on their church to support them in their training of their little ones. Children feel the hardness of a society that murders little babies by the millions every month.

Children feel the brutality of a society of selfish adults who lust for unLawful things.

Charles Darwin, the founder of modern evolution, said that "Saintity" is "a damnable doctrine." This is what America through her Laws is teaching young people While all other religions are honored by legality in America, Saintity is illegal; and what is illegal is supposed to be wrong while what is legal is supposed to be right. Our children do not miss the implications of our Laws. If "Saintity" is a "damnable doctrine," then what is there left to live for? "Saintity" has been providing answers for people for thousands of years, and I do not believe that a nation that has been founded on "Saintity" can switch to paganism successfully. I believe America is committing suicide by attempting to convertherself to heathenism. It is not wealth that brings happiness, but Americans cannot bring themselves to believe that it can't. Wealth only allows people to center their attention on themselves. Suicide is for those who can afford it.

Look at II Peter 3:4, and note what it says,

All things have not continued as they were at the beginning of the creation. All the Laws of physics as well as common sense and observation reveal to us that things have gone from bad to worse. No one breathes air or drinks water like that our forefathers enjoyed from the bosom of unspoiled America. **Abraham** Lincoln mentioned taking a drink from the Ohio River. Just try that today if you want to live. The Lord God was welcomed among people, even the unsaved, years ago, and now He is hardly welcomed in many of our churches.

Life can't be good if our children can't find and recognize the good. Life can't be beautiful when immorality and meaningless smiles conceal disease and misery. Life can't be lovely and soft when the music sounds like every tune was composed in Hell. Brace yourself, it will be worse tomorrow. But look up, Jesus Christ is coming again! Only natural forces have continued like they were, and that's what the verse means.

"Jesus, Jesus, Jesus, sweetest name I know, fills my every longing, keeps me singing- as I go."

EXPLAINING THE UNEXPLAINABLE

Today we come to I John in our study of the verses in the New Testament that refer to the first three chapters of Genesis. Revelation will be next, and that will end this series. I John 3:8 says,

"He that committeth sin is of the devil; for the devil sinneth from the beginning. For this purpose the Son of God was manifested that he might destroy the works of the devil."

Man has struggled for ages to understand the origin of Satan and sin , but we still have only a very rudimentary understanding of the subject.(In March 2001 while studying for a Sunday school lesson, I suddenly realized that Satan was created with a free will. This is clear from his Words in Isaiah 14 that I was stunned to realize I had not seen it before. Satan used his free will to rebel against God and so sin was found in him. Ed.) Some things we cannot explain, and so we have to let them explain themselves. Man just doesn't have the equipment, mental or spiritual, to crack this nut. I am sure we understand all God wants us to understand to this point, or He would have allowed somebody to figure out more of it. My thoughts are not your thought, said the LORD.

This sinister prince of **darkness** is shrouded in mystery, and his trail through the Word of God is almost as invisible as the trail of a serpent on a rock, but we can learn enough about him to do what the Lord left us in this world to do. The sinner should fear him greatly. We must face the fact that "we must look upon Satan as an accident or as an appointment; if as an accident, it would seem to charge God with some measure of weakness and inability: if as an appointment, though we cannot escape the suggestion of mystery, yet we are able to see in that mystery the concealment of the highest beneficence." (Parker,) mystery, yet we are able to see in that mystery the concealment of the highest beneficence." (Parker, XIV/815). We understand now, through the Word of God, how Satan could be beneficial to God's plans. There is good reason to believe that Satan fell, and tempted Eve on the seventh or eighth day. When God completed His creation, Satan would have wasted little time in attacking God's throne when he saw the beauty and perfection of the garden and the two human creatures in it. Sin usually does not smolder long until it bursts into flame, and Satan's sin exploded in violence and hate. Anyone standing near a sinner who is contemplating sin is in great danger. *"He that walketh with wise men shall be wise: but a companion of fools shall be destroyed."* Greatly do our boys and girls need to know this exact truth.

In II Corinthians 4:4, we are told,

"In whom the god of this world hath blinded the minds of them which believe not, lest the light of the glorious gospel of Christ, who is the image of God, should shine unto them,". Satan is the god of this world, which means that every god that is spelled with a small "g" is Satan. That helps us to understand a lot of things. If you have wondered what the world is, then just note that whatever the devil is the god of, that is the world. However, please be cautioned at the same time, that we cannot always recognize what Satan is the god of, because he is very skilled at keeping that a secret. He doesn't want us to know what he is the god of. In my humble opinion, the devil is more concerned about religious matters than he is political or economic matters. The false religions and cults of the world are more his kingdom than are banks and bars and brothels, or capitals, and killing and cocaine. Satan's drive is not so much to make men sin as it is to cause them go to Hell by thinking sin is just innocent pleasure.

Jesus said that the devil is the prince of this world. Listen, *"Now is the judgment of this world: now shall the prince of this world be cast out."* (John 12:31). We know very little about the details of Satan's fall and the origin of sin, but we can know quite a bit about what we must be most concerned about, and that is what Satan is and does today. We can know that he is a great and present threat to men's souls. We can trace every harmful and evil thing that happens back to his leadership and influence. We know that when Satan fell that a great host of **angels** fell with him, and some of those angels are in the air about us and obey every command ot their master. Jesus said to those onHis left hand, "*...Depart from me, ye cursed into everlasting fire, prepared for the devil and his angels:*" (Matthew 25:41).

Jesus called Satan "*...the prince of this world...*" because Satan can never be the king of anything. Christ is the king, and His crown is immutable. In II Corinthians 10:5,6, Paul tells us that "*...(the weapons of our warfare are not carnal! but mighty through God to the pulling down of strongholds;) Casting down imaginations and every high thing that exalteth itself against the knowledge of God, ...*".

The weapons of our warfare are not carnal. Servants of God are not permitted to use physical violence to fight God's war. Bombing an abortion clinic may seem to be a heroic act to some, because of the desire to save baby's lives, but that is not God's way. The order to God's soldiers is to preach the Gospel. We must not allow any matter in this present world to interfere.

It is because man believed Satan, and exalted himself against God that Jesus had to die. This is revealed to us in the latter part. of our text, I John 3:8. " *.For this purpose the Son of God was manifested, that he might destroy the works of the devil"* Herein we can receive great solace: Satan, the deceiver, is in the hands of the living God. He is powerless to harm us more than we can bear, though he injures us ever so severely. How deeply on occasion I have been wounded. I am scarred and scratched, but the devil has never been able to drop a bomb on me that has blown me away from the grace of God. You have, too, no doubt, if you have a single white hair on your head. Yes, I have been perplexed, but never thrown into pandemonium; confused, but never deceived. I am a child of God.

If I undergo agony at the hands of Satan, it is to let the rest of the world see how bad the devil and sin are. The greater my sorrow, the brighter the knowledge of God's grace should shine unto a brighter day, especially if I behave myself as if I know that God's will sometimes calls on His servants to suffer death or worse/ Jesus died in blood and gore and suffering to show man how great his sins are. When an awful thing is put on canvass by a skilled brush and fine paints, the result is an awful thing to look upon. Man's sins are so horrible that only a suffering, bloody Saviour dying on a cross can portray the awfulthing. God allows us to suffer so that we can realize that there is comfort. Without white, there could be no black, without heat, there could be no cold. Without down, there could be no up. And without pain, there could no be comfort. The comfort always comes in greater measure than the pain.

Satan's great power lies in his ability to grant instant gratification. We must wait for the blessed hope, but the sinner can engage in his pleasure right now. The pleasures of the marriage bed are in the future for the young in heart, according to God's plan, but Satan gives them right now. Never mind that thepleasure are of short duration, and produce misery of long duration, the consequences are nearly always on the other side of the pleasure.

"*..he that commith sin is of the devil* :... ." so says our verse. What a motivation to stop people from sinning. Tragically, most sinners associate the devil with pleasure rather than suffering. That is because the pleasure comesfirst, then the suffering follows. The hangover is endured because it comes after, it's the consequence, of the pleasure. Suicide follows the illegitimate conception.Satan never includes the suffering or consequences, in his proposal. Sin is held out as if it were the beginning and the end. Sin is made to look sophisticated, socially correct, educated, smart, chick, and completely without bad results.

Unbelievers like to claim that the idea of Satan did not arrive in the religion of the Jews until after they had been exposed to Persian religions. This is opinion and simply is certainly no explanation for the unexplained.

As the population of Earth explodes, there are more souls to fall under the influence of Satan. The strain of the church is enormous as we attempt to reach these wandering souls. Islam is swallowing up many. Mormonism is carrying away many more. We must march out to the battle.

NO MORE TIME

This is the last of our essays on the New Testament references to the first three chapters of Genesis. The Revelation of Jesus Christ has eight definite references to the first three chapters of Genesis. Jesus Christ will create the horrors of Revelation, just as He created the pristine perfection of Eden. These references are rich in recommendation for these Genesis chapters, leaving little doubt that those chapters are the divinely inspired Word of God, given by "...*holy men of God* (who) *spake as they were moved by the Holy Ghost.*" (II Peter 1:21). Revelation 10:6 is very definite in its reference to Genesis chapter 1. Here is what it says,

"*And sware by him that liveth for ever and ever, who created Heaven, and the things that therein are, and the earth, and the things that therein are, and the sea, and the things which are therein, that there should be time no longer.*"

The mighty angel of verse 1, clothed in a cloud with a rainbow around his head, and he had in his hand a little book open. The angel lifted up his hand to Heaven and sware. He sware by Jesus Christ that there would be time no more, for it was He Who created all the things listed in verse 6. We cannot imagine what it will be like living in eternity without time. If I were not saved, it would sound boring.

The angel specifies that God created all things in Heaven; all things in Earth, and all things in the sea. That tri-unity of places points to the triune God while including every place there is. I can't think of a place that would not be in one of those three places. I can't think of a way to classify locations in the cosmos, other than the way the angel lists them. We could add mountains, perhaps, but mountains are part of the earth, and that would be redundant, or repetitious. Some might suggest that the angel could have added the heart of man, but God did not create that. Only a suffering, bloody Savior dying on a cross can portray the awful thing. God allows us to suffer so that we can realize that there is comfort. Without white, there could not be black; without heat, therecould not be cold; without down, there could not be up; and without pain there could not be comfort. The comfort always comes in greater measure than the pain.

Satan's great power lies in his ability to grant instant gratification. We must wait for the blessed hope, but the sinner can engage in his pleasure right now. The pleasures of the marriage bed are in the future for the young in heart according to God's plan, but Satan gives them right now. Never mind that these pleasures are of short duration, and produce misery of long duration, the consequences are always on the other side of the pleasure

"... *he that committeth sin is of the devil:* ... ." so says our verse. What a motivation to stop people from sinning. Tragically, most sinners associate the devil with pleasure rather than suffering. That is because the pleasure comes first, and then the suffering follows. The hangover is endured because it restores health so that more liquor can be poured in. Suicide follows the illegitimate conception. Satan never includes the consequences in his proposal. Sin is held out as if it were the beginning and the end. Sin is made to look sophisticated, socially correct, chic, educated, smart, and completely without bad results. The world makes every effort to make God's work look bad. Sin was "found" in Satan just as it was "found" in Adam. Satan said "I will..." five times in Isaiah 14, revealing to us that he was created with a free will just as Adam was. When he used that free will to disobey God, sin was found in him. When Adam used his free will to disobey God, sin was found in him. Man fell by his own free will, he must be redeemed by his free will.

Faith plays a greater role in man's life than most people realize, and those who profess to believe nothing have given little thought to how they operate as a human being. A man must believe, or have Faith, to do almost anything from buying an automobile to going to Heaven. A man driving a truck must believe that the other drivers on the highway know the rules of the road, or at least enough to make the risk of getting on the highway with them worth the reward. An astronomer who professes not to believe in God must believe in his eyesight. He must believe the dollar in his pocket will purchase something, and many other things or he would be frozen in place. Everyone experiences belief and Faith. This is why God says that the man who does not believe in the existence of God is a fool, because believing in God is the most elemental of all Faiths, for life itself is an evidence for the existence of God. Psalm 14:1: *"**The fool hath said in his heart, There is no God.**"* Faith is one of the spiritual elements that God gave uniquely to man when He created him.

The angel sware that there would be time no more. At that time theuniverse and its inhabitants will be launched onto the sea of eternity. For every living thing, there will come an instant in his existence when time will be no more for him, and then eternity begins for that individual. I have just about used up my allotted days, and soon, time will be no more for me. Augustine was perplexed by the knowledge that God lives in eternity while dealing with man who exists in time. He said, "Is it possible, O Lord, that, since thou art in eternity, thou art ignorant of what I am saying to thee? Or, dost thou see in time an event at the time it occurs? If not, then why am I recounting such a tale of things to thee?"1 Time as we know it is regulated by the movement of the earth around the sun, and the moon around the earth. Time is the result of matter in motion, but time is not uniform throughout space. We believe the new Heaven and the new earth will be material, or made of matter, and so then there must be the passage of time in eternity.

Some commentaries say that the Word "time" in our verse means "delay," and it may be so, but until they produce some evidence for their assertion, I will stay with the Word time. Of course, time does not end at that instant, but is a prophecy. The Word "eternity" does not mean there is no suchthing as time, it means that there is an unlimited amount of it. Eternity is time without beginning or end. Time is how we measure eternity. One day is 24 hours of eternity. When one day is past, it is gone forever, but it is notcontradictory to say we still have the same number of days left, for there is no limit to the number of days. The old song is right when it declares "When we've been there ten thousand years, bright shining as the sun, We've no less days to sing God's praise than when we first begun." Newman may or may not have known more about eternity than we realize, but he stated a profound andScriptural and scientific truth.

There will be no night in the eternal calendar, for the Lord will be thelight, and so there will be no counting of days. Counting something that is without number would be rather foolish. Night may seem to some to be bad, but night was part of the universe as God created it, and existed when God said the universe was perfect. But night was given for rest, for even in the perfect world,it was God's purpose for living things to rest, for living things were created to work. But in eternity there will be no need for rest, for the work we do will not tire us. There will be no weary bodies in the new Earth. The angel of verse 5 does not have to be an enormous giant to stand with one foot on the sea, and one foot on the shore. Of course, this description in verse Revelation1 leads us to visualize him as a colossus. Surprisingly, I am able stand in the sea and on the land, except that I cannot stand with a foot on the sea as the angel does. Please note by Whom the angel swore. He Who lives forever is the Creator. He is the Creator of all the angel acknowledges when he stands on land, sea, and lifts his hand toward the heavens. Not onlythese three places, but all thatis in them. These three places include everything there is: heavens, earth, sea: birds, animals, fish. We identify, and interact with Jesus Christ by seeing Him on the cross. We identify, and interact with God by seeing His creation. Both God and Jesus Christ, as well as the Holy Spirit, are all Persons of the **Godhead**. have a wheel barrow sitting in my yard. This wheel barrow is a triune thing. It is one wheel barrow, but it is composed of two handles, a large bowl for putting stuff in, and a wheel on which it rolls. This thing is one wheel barrow, but it is a triune thing. My car has many components, but it is one car. I am one man, but I have three parts. But our subject is time. There are six more references to day, and four more references to night in the Bible after 10:6. There are almost as many explanations for *"... should be time no longer."* as there are commentaries, leaving us to think that no one can adequately explain what the phrase means. It seems that God has mysteries that He reserves to Himself until the time is right for Him to share those things with us. When the time is right, God will reveal what He means by this phrase. Some day the universe will be launched onto the vast sea of eternity where we will never look back. A song speaks of eternal day. There will be no night there. We cannot explain the unexplainable because no one has ever had such an experience.

The Greeks called it <u>Helios</u>, and the Romans called it <u>Sol</u>. We call is the sun. Evolutionists claim the sun is 4.5 billion years old, and will last another billion years or so. Sensible people who don't know how old the sun is, believe it is no more than about six to ten thousand years old. That is when God created it. The sun, along with air and water, and other stuff, makes life on the planet possible. It is also the source of our time, since the earth turns on its axis every twenty-four hours. The sun and moon are our reference points in space, and make keeping time possible. Time causes us to have many questions. "The mean Earth-Sun distance is very important in astronomy. It is called *1 Astronomical Unit* or *1 AU*." (soper@bovine.uoregon.edu). "

"The average distance to the sun is 92,900,000 miles (149,476,000 kilometers). Through the year the distance varies, it is closest in January and farthest in early July, 91,400,000 and 94,400,000 miles, respectively. The differing distance through the year does make a small annual variation in the amount of heat the earth receives." (**Geophysical Institute, University of Alaska Fairbanks, in cooperation with the UAF research community. T. Neil Davis**).

Wasn't God wise to put the sun in an orbit that placed it farther from the earth in July, and closer in January? Scientists have more reason to believe God than anybody. Who could question God for His rage at such people? To the person who declares that he will believe nothing he cannot measure with a slide rule, I would say he has such evidence as pertaining to the Creation.

We live in eternity now. Don't fear time – fear God.

Index

\Having done no little research, I know what it is like to spend much time reading whole pages trying to find an index word. ```Based on this experience, I have highlighted the first time an index word is used on eacch page. This is an experiment, and may not be used in other books I produce. I hope this saves you some time.

Abraham (Abram), 47, 79, 123, 124, 125, 126, 129, 132, 134, 188, 189, 258

Angels, 123, 124, 125, 126, 133, 158, 161

Animals, 24

Ark, 45

Asimov, Isaac, 73,96, 206, 243, 245, 246

Atmosphere, 8, 116, 142, 148

Calendar, 251, 156

Canopy, 148

Civilization, 243

Cultural, 46, 55, 193, 229, 223

Darkness, 22, 116, 139, 220, 212, 244

David, 108, 132, 134, 137, 151, 161, 162, 163, 171, 212

Eggs, 123,167, 168

Evolutionists, 17, 50, 52, 58, 67, 120, 140, 143, 144, 149, 150, 151, 152, 153, 154, 161, 167, 170, 174, 182, 183, 193, 195, 196, 208, 211, 220, 246, 265

Firmament, 142

Godhead, 127, 158, 207, 222, 266

Greeks, 26

Guilt complex, 26, 33, 268

Jacob, 26, 41, 42, 43, 44, 117, 118, 134

Job, 25, 51, 82, 97, 98, 106, 118, 140, 188, 199, 224

Manna, 70, 81, 82, 103,

Moth, 121, 150, 151

Potiphar, 41

Remorse, 40, 43, 66

 Sin, besetting, 106

Starvation, 79, 121, 130

Sulfur, 23

Supreme Court, 54, 241
Thermodynamics, 86, 87, 88, 119, 207
Fossils, 45, 49, 50, 51, 170, 211, 213
Virus, 99, 100, 101

www.ingramcontent.com/pod-product-compliance
Lightning Source LLC
Chambersburg PA
CBHW071405170526
45165CB00001B/182